Pioneers of Ecology

by Donald W. Cox

portrait illustrations
by Ted Lewin

Acknowledgments

The Editors are grateful to the following organizations and individuals for their invaluable assistance in the picture research for this work:

American Museum of Natural History: Helen B. Jones
National Audubon Society: Robert C. Boardman, Samuel Dasher
New-York Historical Society: Wilson G. Duprey
New York Public Library: Maude Cole
Theodore Roosevelt Association: Helen MacLachlan
Sierra Club: Daniel B. Gridley

The publisher also makes grateful acknowledgement for permision to use material from the following books:

Beyond the Hundredth Meridian by Wallace E. Stegner. Copyright © 1953, 1954 by Wallace E. Stegner, Houghton Mifflin Co.
George Washington Carver by Rackham Holt. Copyright © 1943, 1963 by Doubleday & Company, Inc. Reprinted by permission of Doubleday & Company, Inc.
Population, Resources and Environment by Paul R. and Anne H. Ehrlich. Copyright © 1970 by W. H. Freeman and Company.
The Quiet Crisis by Stewart L. Udall. Copyright © 1963 by Stewart L. Udall. Reprinted by permission of Holt, Rinehart and Winston, Inc.

ENTIRE CONTENTS ©COPYRIGHT 1971
BY HAMMOND INCORPORATED

All rights reserved. No part of this book may be reproduced or utilized in any form or by any means, electronic or mechanical, including photocopying, recording or by any information storage or retrieval system, without permission in writing from the Publisher.

LIBRARY OF CONGRESS CATALOG CARD NUMBER 77-158132
STANDARD BOOK NUMBERS:
TRADE EDITION 0-8437-3832-4
LIBRARY EDITION 0-8437-3912-6
PRINTED IN THE UNITED STATES OF AMERICA

Contents

Chapter

1	The Ecological Revolution	8
2	Alexander Wilson	14
3	John James Audubon	19
4	Henry David Thoreau	25
5	George Perkins Marsh	30
6	John Wesley Powell	36
7	Guy Bradley	40
8	John Muir	44
9	John Burroughs	50
10	Theodore Roosevelt	54
11	Gifford Pinchot	60
12	George Washington Carver	65
13	Rachel Carson	70
14	Joseph Wood Krutch	76
15	Ian McHarg	80
16	Paul Ehrlich	84
	Environmental Organizations and Periodicals	90
	Selected Bibliography	91
	Index	92
	Credits	93

1
The Ecological Revolution

*God has lent us the earth for our life.
It is a great entail.
It belongs as much to those who follow us
 as it does to us
And we have no right, by anything we may do
 or neglect to do,
To involve them in unnecessary penalties,
Or to deprive them of the benefit
Which we have in our power to bequeath.*

JOHN RUSKIN
1819-1900

What It's All About

Before Earth Day dawned on the American scene in April 1970 the word "ecology" was rarely heard by the average citizen. Only those in the universities or those who were members of our half dozen or so conservationist organizations and readers of their journals understood the meaning and significance of this little-known term. Now, however, the word is on everyone's lips as new headlines flash across our newspapers and TV screens every day concerning the spilling of oil into our streams, lakes and oceans, the dumping of poisonous chemicals into the sea, or a smog blanket creeping over some unsuspecting city, threatening the human life below.

Ecology is a relatively new science, having come to this country some seventy years ago from Europe. It means, literally, the relationship of an organism to its total environment. When one or more of these organisms is out of balance with the surrounding environment, then we have an ecological problem to solve. This is the dilemma facing modern man, as more and more of his technological creations have created poisonous by-products that now appear to do more harm than good to the environment. We produce new, sophisticated chemical compounds for so-called "constructive" purposes to aid mankind, i.e., the refining of leaded and unleaded gasoline to burn in our internal combustion automobile engines to make it easier for us to move about in work and play. Yet at the same time, we have created a monster, which emits into the air harmful deadly hydrocarbons as well as poisonous carbon monoxide and oxides of nitrogen and sulfur.

It is the control of these destructive side effects before it is too late for us all that has most concerned the modern day ecologist. Today's leading environmental advocates, from scientists to politicians, are attempting to make a fundamental point — that man's environment has become so complex and interrelated that any action that alters one aspect of the environment may have a potentially disastrous impact on man's future.

With pesticides, the mechanism is the food chain: plants absorb DDT, are eaten by animals, which in turn are eaten by humans and other animals. Until 1970 American babies were consuming four times the DDT content allowed in the interstate shipment of cow's milk.

America is daily fouling its air, streams, lakes, marshes and the oceans which surround her. We are burying ourselves under a deluge of 8 million scrapped cars, 30 million tons of waste paper, 48 billion discarded cans and 28 billion bottles and jars each year. A million tons of human-generated garbage pile up each day. The air that we breathe—and pollute—circles the earth forty times a year, and America alone contributes 140 million tons of pollutants: 90 million from cars alone (we burn more gasoline than the rest of the world combined) and 15 million from electric-power generation. Noise, straining and jarring our lives, doubles in volume every ten years. Just walk through the center of any large city during rush hour or near a skyscraper building project. There are 5,500 Americans born every day, making a total of 100 million more people by the year 2000 than we have now. Already we consume and waste more than any other people on the earth. We flatten our hills, fill our bays, leave ugly, scarred slag heaps where mountains of coal once rested, and blitz our wilderness with gasoline-powered saws. The quality slowly drains from our lives.

Historical Development

For the first three and one-half centuries after Columbus' "discovery" of America, there was little concern among the nation's settlers about conservation. It may be said that an iconoclastic naturalist started it all by taking to the woods outside of Concord, Massachusetts, just prior to the Civil War. What Henry David Thoreau kindled with his back-to-nature experiment at Walden Pond has continued to flower slowly ever since.

In a sense, the stories of the fifteen ecological leaders told in this volume mirror the four major eras of concern for conservation on the part of the American public. The first era was centered around individuals like Thoreau, John James Audubon and George Marsh, who, among thoughtful citizens, awakened respect for wildlife and the desire to protect it. No government action or deep national concern accompanied these small beginnings of the movement in the mid-nineteenth century.

The second era of ecological concern began at the end of the century. With the closing of the frontier in 1890, conservation publicists like John Muir and John Burroughs spurred President Theodore Roosevelt and his chief forester, Gifford Pinchot, to use their political power to preserve for the coming generations the wilderness wonders that Roosevelt had known in his youth. They would preserve, "conserve" it all — the unspoiled skies, clear streams and wildlife resources that were vanishing from the land.

The third era of conservation dawned a generation later in the wake of the Great Depression of the thirties. Man's economic plight was compounded by the winds that created the dust bowl in the Southwest and the uncontrolled floods in the Mississippi, Tennessee and Ohio River valleys. The land's inability to hold back the water as a result of

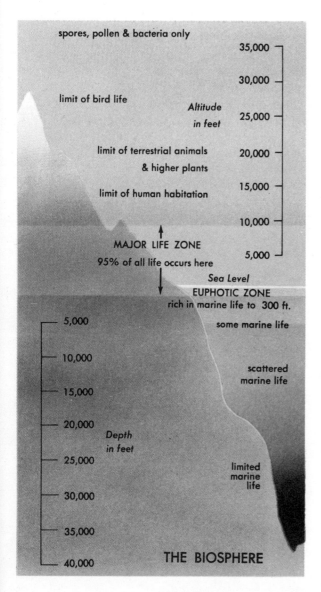

A thin film of air, earth and water around the globe contains the living organisms of our planet. Within this 'life zone' occur all the energy transfers, chemical cycles and processes of interaction which sustain our major activities. The biosphere, as this zone is known, is only twelve miles thick. In fact, within an even narrower band only two miles thick can be found 95 percent of all life.

being stripped of its protective cover caused President Franklin D. Roosevelt to take drastic action. Under crusading Secretary of the Interior Harold Ickes, the federal government established the Soil Conservation Service, set up tree-belt windbreaks, formed the Civilian Conservation Corps and the Tennessee Valley Authority as an antidote to the rape and neglect of the land and rivers of America.

Unfortunately, this noble effort was largely halted by the onset of World War II. The TVA still remains, however, as a shining example of what can be accomplished constructively by the government, since its record of the prevention of floods in that once ravaged basin has erased that scourge forever. When the war was over in 1945, government concern was not revived. For a generation there was no one working to protect the environment except for a few loyal believers in such conservation-minded organizations as the Audubon Society, the National Wildlife Federation and the Sierra Club.

During this period America quietly doubled its population and industrial production while tripling its output of automobiles. The resulting blanket of smog and poisonous fumes that hung over our cities and highways began to cause concern among some people.

Oysters and blue-claw crabs were vanishing from our bays and coves. The water in our lakes and streams was rapidly becoming undrinkable, unswimmable and unfishable as we poisoned them with refuse, organic wastes, industrial pollutants and raw sewage. Lake Erie became the classic case. By 1969 it was pronounced "dead," killed by man. "The two natural containers of the environment, the air and water," wrote Theodore White, a noted political scientist and national observer, "finally vomited back on Americans the filths they could no longer absorb."

Man had finally joined the gorilla as the only other animal on earth which fouled its own nest. But people who said that in 1960 were dismissed as "kooks."

The late Rachel Carson helped the cause of public awareness with the publication of her classic, *Silent Spring*, in 1962. The clarion cry was heard and the debate began in earnest. The federal government responded with directives and anti-pollution laws aimed at specific problems. A new era of national concern had begun.

"When we came in, in 1960," observed Stewart Udall, the crew-cut former secretary of the interior and chief environmentalist in the Kennedy-Johnson administrations, "not a single new national park had been set aside since 1947, and all but five percent of the country's free coastline was shut off. The Eisenhower administration had thought pollution was a local matter, so we all sat there like spectators and watched Los Angeles wrestling with smog—it was *their* problem. I came in as a classic conservationist—you know, preservation of nature and seashores, or birdlife and wildlife, of endangered species. Then gradually it came over me that man himself was an endangered species, that we were part of the same chain of life as the birds. Only in the last three years I was in office (1965-1968) did I see it as a whole piece. We'd erred in thinking environment was simply a matter of managing natural resources. What had to be managed was man himself."

In one of the rare moments when politics and history cross, President Nixon, during the latter months of 1970, set in motion a revolution in the government, reorganizing the whole federal structure so that he could properly deal with the ecological miseries of this country. What he accomplished with no controversy was to reshuffle eighty-four different federal agencies, wallowing in five competing departments, into a more relevant federal structure.

Backed by presidential task force recommendations, he set up two new super agencies, the first being the Environmental Protection Agency (EPA) which would monitor and regulate man's everyday actions with respect to the thin membrane of life which we call our environment. Air and water control as well as solid waste agencies all were placed under this new umbrella.

The second is the National Oceanic and Atmospheric Administration (NOAA) which was established simultaneously with the EPA to monitor the global container—the entire hollow of the sky, surface of the earth and ocean depths. In this new structure could be found old-line agencies like the Weather Bureau and the Commercial Fisheries Bureau, formerly in other departments.

Congress also has a role to play in the attack on environmental problems. A significant turning point occurred on September 22, 1970, when the U.S. Senate passed unanimously by a vote of 73 to 0 the Clean Air Bill. This bill requires that auto companies come up with a drastically modified car engine by January 1, 1975, that would reduce car exhaust emissions by ninety percent or face heavy fines and plant closings. Despite the complaints from the auto magnates that the job couldn't be done within the four-years-plus deadline, this move on the part of Congress signaled the realization that something drastic must be done, and done soon, before we all suffocate from auto fumes.

If we could put a man on the moon before the decade of the sixties was ended—thus fulfilling the late President Kennedy's deadline by achieving an impossible space transportation goal in eight years at a cost of $30 billion—surely we can, in half that time, reduce auto pollution at a fraction of that cost, since the technology necessary to accomplish the land transportation goal is not as complex.

We would also be making a serious mistake if we abdicated our responsibilities on the local and state governmental levels and left the task entirely in the hands of the federal government. An example in the area of growing local concern occurred in November 1970 when Suffolk County, Long Island, banned the sale of phosphate detergents in the local supermarkets. This drastic action took place after the drinking water became polluted and residents rose up in arms when soapsuds started pouring out of their water spigots. When pollution struck close to home, the citizenry became ecologically concerned overnight and brought public pressure on their local government to act fast.

The city of San Diego presents a positive example of what can be done on the local level when a community gets together with a will to rectify a wrong to nature. The once-beautiful bay was so polluted by 1960 that the sardines and other commercial fish disappeared from its waters. With the adoption of a new sewage disposal system by the city and strict controls on industrial waste disposal, the waters slowly returned to normal. By the end of 1970 the sardines had returned.

It is already obvious that the fourth era in the conservation movement is still far from over. The ecological revolution bids well to become one of the underlying issues of the 1970s. Reorganization of government regulatory and operating agencies in the area of conservation is not enough. Even if we ultimately establish a new Department of Natural Resources as proposed and enlarge the authority of the federal government even more than has already been done, this will not alone solve the problem.

Only a united effort, employing the cooperation of all of the people in this nation and on the whole planet, can accomplish the elimination of environmental pollution. As the late Rachel Carson put it so well in her last major speech when she received the coveted Audubon Medal in December 1963:

"We are all united in a common cause. It is a proud cause, which we may serve secure in the knowledge that the earth will be better for our efforts. It is a cause that has no end: there is no point at which we shall say, 'Our work is finished.'

"It inspires hope and courage because even the darkest hours have brought forth men who, though their numbers were but a handful, nevertheless contrived to save for us enough that we may at least glimpse the grandeur of this continent as it was a few centuries ago.

"If the crisis that now confronts us is even more urgent than those of the early years of the century—and I believe it is—this is because of wholly new factors peculiar to our own time . . . .

". . . we live in a time when change comes rapidly—a time when much of that change is, at least for long periods, irrevocable. This is what makes our own task so urgent. It is not often that a generation is challenged, as we today are challenged. For what we fail to do—what we let go by default—can perhaps never be done.

"I take courage, however, in the fact that the conservation effort has a broader base than ever before. There is more organized effort; there are many more individuals who are conscious of conservation problems and who are striving, in their own communities or on the national scene, to solve these problems . . . ."

The current ecological revolution has given all educational and age levels in our nation a chance to work cooperatively on joint solutions that affect us all, rich and poor. Because we are all now concerned with sustaining our own lives and those of our children, the issue of ecology is no longer controversial. What is controversial is *how* we are to attack the problems of pollution, overpopulation and limitless ingestion of limited resources, *who* is going to organize the fight, and *where* the money is going to come from to pay for the enormous costs of cleaning up our air, water and landscape.

There is hope now that we can soon bury our other differences of race, religion and politics in order that we might get on with the task before us. The challenge of the opposing poles of destruction and survival beckons us.

Spaceship Earth, Will It Survive?

The late Adlai Stevenson warned us in his last speech that "we travel together, passengers on a little spaceship, dependent on the vulnerable supplies of air and soil. . . preserved from annihilation only by the care, the work, and I will say, the love, we give our fragile craft."

But, unlike the man-made spaceships that we have sent to the moon and beyond the earth's protective atmosphere, all our life-support systems are right here on this planet. They are not monitored elsewhere, watched over and guarded while we are asleep, as our Project Apollo astronauts are watched over via hundreds of consoles at the Manned Spacecraft Center in Houston.

If mankind insists on sleeping while the water, air, greenery, animal, fish and birdlife become polluted, destroyed or caused to disappear, no human or mechanical monitors are going to flash warning signals to alert us to impending disaster. . . unless we prepare for such a series of events now.

If there is a common thread to the lives of the American ecological pioneers profiled in the following pages, it is this: that each, in his own way, tried to make this country a better place in which to live by helping to save man from his greatest enemy—himself.

2
Alexander Wilson

Alexander Wilson was a mercurial and enigmatic poet, teacher and naturalist who became our first great American ornithologist. He was the first man to study and paint the birds of North America in their natural surroundings.

It is not a coincidence that the first true conservationists and environmental ecologists were bird lovers. Men like Wilson and, later on, Audubon, though they are primarily remembered as ornithologists, were also champions of the preservation of the habitats of winged wild fowl. The Audubon Society — America's first group of conservation-minded citizens — started out purely as preservers of birds but has since broadened its scope to include all branches of conservation of natural resources.

Born in the bleak, wild and romantic town of Paisley, Scotland, in 1766, the son of a weaver and sometime smuggler, young Wilson was apprenticed to a weaver. He soon rebelled and became a peddler of silks and muslins. While traveling through Scotland he started to write and publish poetry, which he sold along with his more material products. One of his popular poems, *Watty & Meg,* written in a dialect style similar to that of Robert Burns, sold more than 100,000 copies in a couple of weeks.

When a labor controversy arose between the weavers and millowner overseers in Paisley, young Wilson took up his pen in defense of the oppressed weavers and wrote a satirical verse about the manufacturers. The times were tense because of the impact of the French Revolution, and Wilson's sardonic poems caused the ex-weaver to be fined and thrown into jail. As a lasting insult he was forced by the British authorities to burn his satirical writings on the steps of the town jail.

He was also accused of blackmail, which, although never proven, led him, embittered and in poverty, to flee to America. He landed at New Castle, Delaware, in July 1794 and walked the few miles north to Philadelphia, where he obtained a series of jobs as a peddler, weaver and teacher in southern New Jersey and eastern Pennsylvania.

While in the City of Brotherly Love, he became fascinated with the birds which he saw there, since they were so much more colorful than those in his native Scotland. For a while he taught at a little stone school in Gray's Ferry, then a small suburb in southwest Philadelphia. Nearby, on the banks of the Schuylkill River, stood the John Bartram house, home of America's first botanist family. William Bartram, though in his sixties, struck up a lasting friendship with Wilson, who was then just twenty-eight. The saintly old Quaker naturalist soon discovered that the moody young Scottish emigré was living too much within himself, and so he encouraged Wilson to draw and paint, giving him prints of flowers to copy.

Wilson soon was busily sketching mammals, flowers and birds by candlelight after his daytime teaching chores were finished. He filled his room with live opossums, squirrels, snakes, lizards and birds. Even his eager pupils brought him live specimens — mice, frogs, and a basketful of crows — to add to his menagerie and to provide subjects for his sketch pad.

In a letter to Bartram expressing appreciation to his teacher, Wilson wrote: "... I confess that I was always an enthusiast in my admiration of the rural scenery of Nature: but, since your example and encouragement have set me to attempt to imitate her productions, I see new beauties in every bird, plant or flower I contemplate."

Gradually Wilson's interests narrowed to birds, and in 1803 he formulated a plan in his mind to produce an illustrated volume on the subject titled *American Ornithology.* He knew the names of only a few of the birds

that he sketched, but with the aid of Bartram, who had listed some 215 different species in his book, *Travels,* Wilson soon became well acquainted with the birds in his immediate area.

In 1806 he resigned his teaching post to accept an assistant editorship with *Ree's Encyclopedia,* which was in the process of being published by Samuel Bradford in Philadelphia. Wilson presented his idea to Bradford who agreed to publish the *Ornithology*. Thus encouraged, Wilson used all his free time to work feverishly on his project.

In September 1808 the first volume of *American Ornithology* appeared with the bird plates drawn by Wilson. The quality of the book far surpassed anything of its kind published in America up to that time. The high $120 price for the proposed ten volumes was also a phenomenon for its day.

Shortly after the appearance of the first volume, Wilson set out on a trip through New England in search of "birds and subscribers" for his books. Upon returning to Philadelphia from this trip, he remained only a couple of days before setting out again on a three-month journey by horseback to Savannah, Georgia. By the time he reached his southern destination in March 1809, he had 250 subscribers and had visited "every town of importance within 150 miles of the Atlantic Coast." His travels were costly, however, and when he returned to Philadelphia by boat he found that his meager savings were almost depleted.

In January 1810 Wilson began his longest and most ambitious journey, from Philadelphia to New Orleans, to collect more birds and to increase his list of subscribers. He traveled by stagecoach and on foot to Pittsburgh, then started down the Ohio River in an open skiff. Just after leaving Pittsburgh he took the first of four snowy owls that he observed on the trip.

After the first day on the cold river the ice floes disappeared, and Wilson shared the open water with boats from Kentucky, which were taking supplies and passengers to the West. At night he stayed in cabins and villages dotting the riverbank. Paddling down the river he noticed several great horned owls, kingfishers and cardinals. On March 5, 1810, he got caught in a violent storm where the Scioto River met the Ohio. On that day he saw for the first time a rare flock of Carolina parakeets.

Later on he shot several and wounded one, which he succeeded in taming. This little bird continued on the trip with Wilson and was his sole companion for hundreds of miles.

His remarks on the peculiar characteristics of the parakeet foreshadowed their eventual early extinction by man.

"Having shot down a number," he wrote ... "the whole flock swept repeatedly around their prostrate companions, and again settled on a low tree, within 20 yards of where I stood. At each successive discharge ... the affection of the survivors seems to increase, for, after a few circuits around the place, they again alighted near me ..."

At Louisville Wilson sold his skiff, named the *Ornithologist*. The man who bought it thought Wilson had given his boat a droll Indian name and commented laconically: "Some old chief or warrior I suppose?"

It was in March 1810 that Wilson had his first and historically-famous meeting with John James Audubon. With his tame parakeet on his shoulder and his portfolio of bird drawings under his arm, Wilson entered Audubon's country store in a Louisville suburb and showed the proprietor his accounts of the life history and drawings of birds that he had seen on his travels.

Wilson impressed Audubon with his statement that what he had seen was a preview of the ten-volume *American Ornithology*. "It shall contain the life histories and portraits of *all* the birds of America," he said proudly. Audubon then showed his own portraits of birds to the equally surprised Wilson. Not suspecting that he had a competitor, Wilson asked the young storekeeper if he intended to publish his own work. The idea had not occurred to Audubon, who otherwise might have subscribed to Wilson's book if his own business had not suffered from long hours spent in collecting and painting.

Two days later the two bird fanciers wandered the surrounding Kentucky woods together, but it is doubtful that either took the other's ornithological pursuits too seriously at the time. Strangely, this was the only lengthy meeting of these two men. Although their interests were so much alike, their personalities differed greatly and prevented a mutual sharing of their dual artistic and scientific pursuits.

After leaving Louisville in mid-April Wilson continued his journey overland, where he observed many more birds, including a huge flock of passenger pigeons and their three-by-forty-mile-wide breeding grounds at Danville, Kentucky. He noted that every tree was loaded

Below: The Bartram house near Philadelphia was a second home to Alexander Wilson.

The journeys of Wilson in America.

with pigeon nests, and one had over ninety. He spoke of one flock of these birds taking almost three hours to pass overhead.

As he traveled south through the spring-blossoming woodlands, Wilson spotted warblers, grouse, whip-poor-wills, rails and other feathered species. Beyond Nashville he entered the wilderness of the Indian country, and slept in the huts of the Chickasaws, and was pleased to see the gourds they hung in their camps for martins to nest in. While leading his horse through the swamps and canebrakes of western Tennessee, he contracted a fever and dysentery and nearly died.

Finally on May 18 he reached the Mississippi at Natchez and, while staying at the plantation near there, he discovered and named the Mississippi kite (as he had previously done for the Kentucky, Tennessee and Nashville warblers, which he also discovered).

With his pockets crammed with bird skins, he made his way south to New Orleans, and in seventeen days in the Crescent City, he succeeded in obtaining sixty subscribers to his ornithology before he booked passage on a ship to New York. All told, he obtained 458 subscribers on his canvassing tours — including such well-known people as Thomas Jefferson, Gouverneur Morris, DeWitt Clinton, James Monroe, Benjamin West as well as the most important Ivy League college libraries in the East.

President Jefferson befriended Wilson and Meriwether Lewis, explorer of the American West, gave Wilson specimens of the tanager, crow and woodpecker, from which he made three of his best paintings. In his travels on the Natchez Trace, down the Ohio River, and in Middle America, Wilson catalogued and painted dozens of birds. His precise, candid descriptions are masterpieces of nature writing. His book, *American Ornithology,* remains unrivaled for the clarity of its paintings and the accuracy of its descriptions.

From 1811 to 1812 Wilson lived almost constantly at the Bartram homestead near Philadelphia where, in the quiet of the botanist's garden, he composed six more volumes of the *Ornithology.* During this period he made many short trips to the headwaters of nearby rivers, such as the Schuylkill and the Lehigh, and to the Pocono and Blue Mountain regions of Pennsylvania.

Wilson also made six trips to the Atlantic coast, including several to the Cape May and Great Egg Harbor regions, in the company of George Ord, who completed the *Ornithology* after Wilson's death. The last trip to the shore was made in the early summer of 1813, where he spent four weeks collecting material for his eighth volume on such water birds as the gull-billed tern, fish crow and the northern seaside sparrow.

Finally on August 23, 1813, in his forty-eighth year, Wilson suffered a recurrence of dysentery and died, his work still incomplete. Reduced in his last days to coloring his own plates as a means of livelihood, Wilson ended his life in poverty, but he did not die without honor and recognition. In his last years he lived to be chosen as a member of the Society

Poem by Alexander Wilson written during his trip to Niagara Falls in 1804.

THE

FORESTERS, &c.

SONS of the city! ye whom crowds and noise
Bereave of peace and Nature's rural joys,
And ye who love through woods and wilds to range,
Who see new charms in each successive change;
Come roam with me Columbia's forests through
Where scenes sublime shall meet your wandering view;
Deep shades magnificent, immensely spread;
Lakes, sky-encircled, vast as ocean's bed;
Lone hermit streams that wind through savage woods;
Enormous cataracts swoln with thundering floods;
The settler's*(1) farm with blazing fires o'erspread;
The hunter's cabin and the Indian's shed;
The log-built hamlet, deep in wilds embrac'd;
The awful silence of th' unpeopled waste:
These are the scenes the Muse shall now explore,
Scenes new to song and paths untrod before.

* *For Notes see Appendix.*

A 2

of Artists in America and of the American Philosophical Society.

Wilson was not a great artist, and he borrowed much of his writing technique from Bartram. However, because of his gift for accuracy, Wilson's work became the foundation and set the style for later writings and studies of bird life. He tried to dispel erroneous ideas concerning the destructive feeding habits of birds by examining the contents of the stomachs of the birds that he took, and wherever possible he made a plea for bird conservation.

A forerunner of later conservationists, Wilson wrote on the yellow-billed cuckoo: "...from the circumstances of destroying such numbers of very noxious larvae, they prove themselves the friends of the farmer, and are highly deserving of his protection."

In summing up the arguments of the contributions of the ivory-billed woodpeckers in feeding on insects, Wilson wrote: "Until some effectual preventive or more complete mode of destruction can be devised against these [tree eating] insects, and their larvae, I would humbly suggest the propriety of protecting, and receiving with proper gratitude, the services of this and the whole tribe of woodpeckers."

Wilson's *American Ornithology* contained the portraits of 262 species — of which thirty were new to the budding science. His discoveries included the goshawk, black-billed cuckoo, canvas-back, mourning warbler, song sparrow and long-billed marsh wren — among others. Wilson's classic work, with Audubon's *Ornithological Biographies,* brought the study of birds into the field of American science for both amateur and professional ornithologists. He is remembered not only as the Father of American Ornithology but as one of our pioneer ecologists and conservationists who fought to preserve a portion of the American natural wilderness for posterity.

3
John James Audubon

Why have the exquisite paintings of John James Audubon dominated the art of bird painting around the world for almost a century and a half? The answer lies largely in his uncanny ability to infuse his work with vitality and enthusiasm and because he had a deep understanding and passion for nature. He drew and made watercolor paintings of birds, animals, flowers and trees with an intense concentration and amazing fidelity that has withstood the tests of time.

His original art, which was rendered by using a combination of pencil, watercolor, pastel, crayon, ink, oil and lacquer, foreshadowed the later works of the multi-media artists of the mid-twentieth century. Truly, Audubon was a man many years ahead of his time. His ability to capture the texture and colors of his bird specimens helped to convince his admirers of the need to conserve these beautiful creatures when they were later threatened with extinction by man during the latter part of the nineteenth century.

Audubon's paintings have no counterpart in Western art, but rather a kinship with Chinese and Japanese engravings. Both shared similarities of technical virtuosity and brilliant naturalism, but he outshone Oriental artists with the accuracy of his observations.

John James Audubon was born April 26, 1785, at Les Cayes in what is now Haiti. His father was a French maritime officer and plantation owner. His mother, a Creole of Haiti, known only as Mademoiselle Robin, died shortly after his birth. He was not born, as he later wrote, near New Orleans nor, as others contended, in the French Royal Court. Legend for many years labeled Audubon as the lost Dauphin of France.

During the American Revolution Captain Jean Audubon was for a time prisoner of the British in New York. While engaged in sugar and coffee trade he often came to the United States. During one of his sojourns in America he acquired a farm and home at Mill Grove on the bank of Perkiomen Creek near Philadelphia. Captain Audubon gave up his plantation in the Caribbean when slave insurrections swept the Antilles and in 1789 returned to France with the four-year-old Jean Robin. Four years later the young Audubon was legally adopted by Captain Audubon and his wife.

Audubon's education seems to have been sporadic despite his father's hopes for a naval career. In 1803, with war clouds overhead, the elder Audubon decided to help his son escape conscription into Napoleon's military by sending him to America to learn farming and mine management at Mill Grove.

Young Audubon, who was nineteen when he arrived at Mill Grove, soon lost interest in either farming or operating the lead mine on the property. The lure of the surrounding Pennsylvania wilderness was greater. He devoted most of his time to roaming the wooded hills near his home and to hunting, painting, taxidermy and the courting of Lucy Bakewell, the daughter of a cultured English neighbor.

Lucy, whom he married three years later, became the inspiration of his life. During his eighteen years of poverty and failure in business ventures and his thankless early struggles as a nature artist, she resolutely stood by him. Through her own earnings as a schoolteacher, she freed him to carry out his great plan of painting *all* the birds of America. Without her loyalty and encouragement he might never have seen his life task completed.

Soon after his marriage Audubon and his bride moved to Kentucky where he was to spend the next twelve years attempting to earn a living in a series of business ventures. With a partner, Audubon established a country store in Louisville, but in 1810 the business was moved downriver to Henderson. It was at this time that a historic but brief meeting

with Alexander Wilson, another great ornithologist, took place. Audubon was already spending too much time away from the business with his sketching of birds and nature rambles. Wilson's visit inspired him to work harder at his sketching of birds and the business suffered even more. Another move was made but it too ended in failure.

Bankrupt by 1819, Audubon tried to earn a living doing chalk portraits and teaching art. In 1820 he moved to Cincinnati with his family, which by now included two sons. He worked for a few months as a taxidermist, but his energies were bent more on sketching birds. It was at this time that his drawings began to show considerable improvement, being less stiff and more colorful than heretofore. The idea of publishing an ornithology of North America began to take hold. On October 12, 1820, Audubon left the Ohio River city on a flatboat bound for New Orleans. With him also went his assistant, Joseph Mason, a thirteen-year-old pupil, who was remarkably gifted at painting flowers. Mason supplied the floral background in at least fifty of Audubon's drawings. Audubon would paint the bird first and then select the plant, which he gave to Mason to put into the background.

In the summer of 1821 at Oakley Plantation, some one hundred miles up the Mississippi River from New Orleans, Audubon reached the height of his artistic capacities. There he executed twenty of his best drawings. Employed as a tutor, he was relatively free for the first time from financial worries and could spend at least half of his time drawing birds. Here, far from the distractions of the city, he could work at his craft midst the solitude and stimulation of nature in back-country Louisiana.

Unfortunately, this idyllic situation ended in late summer, and Audubon and Mason reluctantly returned to New Orleans. That

Mill Grove where Audubon first lived after coming to America.

winter Audubon's family joined him. Lucy brought along his earlier drawings, which were of low quality. He thereupon decided to do most of them over again.

When his wife started a private school in nearby West Feliciana Parish in 1823, he became a drawing instructor in her school. In 1824 he met George Lehman, a well-known Swiss watercolorist, who soon became his second assistant and helped Audubon to complete a series of superb paintings over the next several years.

By 1824 he had enough material to consider publication and he took his drawings east in search of a patron. Being unable to find a publisher in either Philadelphia or New York, he sailed for England. After a successful exhibition in London, he found a publisher, Robert Havell, willing to print his classic *Birds of America* on a subscription basis. This work was published in several parts beginning with the first volume in 1827.

For the next twelve years Audubon divided his time between Europe and the United States. He made trips to Florida, Maine, New Brunswick, Labrador and Texas to gather material for the series. Additional assistants were taken on and the Audubon sons, Victor and John, began to help in the preparation of prints.

As he hastened to complete his major work, Audubon became frantic and was forced to making drawings from skins only. In his haste to depict all the new species, he had to crowd many birds on a single plate, so much so that some of the latter of his 435 plates were poorly drawn.

The text to accompany the *Birds of America* was written in collaboration with William MacGillivray, who was responsible for the scientific information about each bird species. Unfortunately Audubon was careless and stubborn about giving due credit to his assistants, which was probably the worst of his faults.

Perhaps unsure of his own great talents, he fabricated a story that he studied painting under Jacques Louis David, a renowned painter at the French Court.

During his life Audubon made at least 1,000 pictures of birds, most of which contain more than two individuals in each drawing or painting; 78 watercolors of mammals plus 100 portraits, landscape sketches and watercolors of eggs. Over the years, he developed from a stiff-stroked artist to a giant craftsman and watercolorist of the first rank. He never tired of painting the same species over and over until his work satisfied his own meticulous standards of excellence.

To keep a freshly killed bird in a natural position for drawing, he developed a technique of inserting a series of wires in the carcass and forcing the ends into a block of soft wood. Since he made his drawings life-size, he was able to check each portion of the bird's anatomy with a pair of dividers so that he could make his work all the more accurate.

As a conservation-minded naturalist, Audubon also painstakingly studied the mannerisms of each species in its live habitat, so that he seldom failed to capture a characteristic pose of that particular bird.

The success of his *Birds of America* enabled Audubon to build a house for himself and his wife on northern Manhattan. Here he continued his work, despite unremitting worries and discouragements, completing a smaller edition of his bird portraits.

In 1843, with the bird books behind him, Audubon took a long-cherished field trip to the Far West but unfortunately did not reach the Rocky Mountains as he had planned. With him went Edward Harris, a wealthy amateur naturalist whom he had met in Philadelphia many years before, the printer Isaac Sprague and two other assistants. Their trip took them up the Missouri River as far as the Yellowstone.

Audubon's travels
through America 1803-1843.

Audubon returned to his home overlooking the Hudson River with many new drawings and specimens for his next major work, *The Viviparous Quadrupeds of North America*. This work, to be a sequel to the ornithological volumes, was done in collaboration with John Bachman of Charleston, South Carolina. The first volume appeared in 1845. A year later Audubon's eyesight began to fail. As his drawing skill declined his two sons, competent artists in their own rights, took over the task of completing the *Quadrupeds*. In 1847 Audubon suffered a serious mental collapse from which he never recovered. He died less than four years later, on January 27, 1851.

Of all the things written about and by Audubon, the two thoughts best representing his interest in ecology were probably those penned by him at two stages in his career.

The first, written in the form of a letter dated October 11, 1829, shortly after leaving the Great Pine Forest, reads: "After all, there is nothing perfect but *primitiveness,* and my efforts at copying nature, like all other things attempted by us mortals, fall far short of the originals. Few better than myself can appreciate this more with despondency than I do."

Nothing, however, sums up his spirit more eloquently than the following advice pub-

Audubon watercolor of Arctic tern probably done in 1835.

lished near the end of his life in the last volume of his *Ornithological Biography* and directed at potential artist-naturalists in the future: "Now, supposing that you are full of ardour, and ready to proceed, allow me to offer you a little advice. Leave nothing to memory, but note down all your observations with *ink,* not with a blackhead pencil; and keep in mind that the more particulars you write at the time, the more you will afterwards recollect. Work not at night, but anticipate the morning dawn, and never think for an instant about the difficulties of ransacking the woods, the shores or the barren grounds, nor be vexed when you have traversed a few hundred miles of country without finding a single new species.

"It may—indeed, if not infrequently, *does*—happen, that after days or even weeks of fruitless search, one enters a grove, or comes upon a pond, or forces his way through the tall grass of a prairie, and suddenly meets with several objects, all new, all beautiful, and perhaps all suited to the palette. Then, how delightful will be your feelings, and how marvelously all fatigue will vanish!"

This vision of the "wonders" of life on the American frontier as expressed by one of our conservationist giants can be rekindled by our generation if we set our minds to it. Audubon in both his writings and art gave us a lasting impression of the beauty of rural America a century and a half ago.

Audubon prophesied the destruction of wildlife by man in America. He believed with a passionate conviction that in years to come no one would ever again have the same opportunity as he to study the birds of America in their haunts. This realization drove him mercilessly in his disciplined and exhausting quest. He saw sights that have all but vanished from the American scene and some that have gone forever—flights of great white whooping cranes majestically winging their

River otter drawn by Audubon for his *Viviparous Quadrupeds of North America*.

way down to the Gulf Coast from their summer stay in Canada; ivory-billed woodpeckers, the largest axmen of their winged tribe, chopping bark from trees to get at grubs and other food; colorful flocks of chattering parakeets; passenger pigeons that gathered together in such astronomical numbers that they blackened the sky in their flight, day-after-day, while filling the air with the deafening thunder of their wings.

Early in his career Audubon took pleasure in shooting birds in order to identify them and obtain the best specimens for his paintings. In this respect, he was far unlike his modern followers who hunt birds only with binoculars and camera. Once, in a single day, he boasted of shooting enough birds in the Florida bayous to make a feathered pile the size of a "small haycock."

In his later years, although he never seemed to regret his own big kills of an earlier day, he changed his attitude and began to lash out at the reckless raids on wildlife. He lamented the disappearance of deer in the East and denounced the "eggers of Labrador," who slaughtered sea birds for fish bait. On witnessing the ruthless actions of one fur company killing large numbers of mink and marten, he cried out in anguish, "Where can I go now, and visit nature undisturbed?"

Stephen Vincent Benét, a contemporary twentieth-century poet and essayist, wrote that Audubon lived to look at birds in all their natural freedom—darting in pursuit of tiny insects, arguing and scolding among themselves, grubbing for larvae in dead trees, taut with panic before their natural enemies, pouncing on their prey, or tranquilly perched amid the leaves and flowers of their natural habitat.

His facility for the observation of birds as expressed in his paintings and drawings is still his chief claim to fame. Indirectly he gave later conservationists and admirers a vision of America when it was still a land where the balance of nature's life cycle worked harmoniously—a situation which we are trying to restore today. His works have been reprinted many times and helped give birth to the Audubon Society, named in his honor, to promote the preservation and study of wildlife.

In 1951, his old 122-acre farm and home, Mill Grove, near Norristown, Pa., was purchased as a living memorial to Audubon in the place where the great ornithologist first came to know and paint the birds and other animals of his adopted land.

President Harry Truman issued a special proclamation on September 18, 1951, making that year an official Audubon Centennial Year since "John James Audubon, naturalist, ornithologist, and artist, by his devotion to a task which he loved and to which he dedicated his life, made an outstanding contribution to American culture and art. . . .

"Audubon was a forerunner of the movement for the conservation of wildlife in America, and his work continued to stimulate appreciation for the wealth and beauty of America's natural resources, serving as a constant inspiration in the continuing endeavor to preserve our birds and other wildlife from extinction. . . ."

No finer tribute could have been made to this pioneer.

4
Henry David Thoreau

On May 6, 1862, when he died at the age of forty-four, Henry David Thoreau was a rather obscure naturalist, lecturer and writer. Today, just over a century later, his name is known on every continent for his contributions to the betterment of mankind, inspiring the modern day ecological-conservationist movement.

Thoreau, an extreme individualist in the sturdy New England Yankee tradition, rejected all the pressures of society to conform to the mores and customs of the day, which has made him an inspiration to people who question today's technological civilization.

Born on July 12, 1817, at his Grandmother Minott's farm just outside Concord, Massachusetts, Thoreau was descended from French Huguenots and Scottish Quakers. The family was quite poor during Henry's early childhood as his father struggled and failed both as a storekeeper and a farmer. In his early years, Thoreau showed a great love for the rural countryside surrounding the village of 2,000 souls. He also showed an aptitude for learning. With the help of his sister, a young school teacher, and others he enrolled at age sixteen at Harvard where he soon excelled as a student, standing near the top of his class. In his third year at Harvard, he was profoundly influenced by Ralph Waldo Emerson's slim volume *Nature* (which had been published in 1836), particularly by such lines as ... "The lover of nature is he whose inward and outward senses are still truly adjusted to each other.... We are taught by great actions that the universe is the property of every individual in it."

After graduating from Harvard where he studied Latin, Greek and the English classics, Thoreau taught school for a while, then turned to lecturing and writing. In 1845, when he was twenty-eight years old, he began his famous Walden Pond "back-to-nature" experiment of simple living, spending two years placidly studying nature while living in a hut of his own building, located just two miles from Concord. He was in no sense a complete recluse during this period, since he grew beans for sale on a few acres, took odd jobs in town as a surveyor, helped in the family's pencil-making business and walked into town to spend evenings with friends, including Ralph Waldo Emerson.

His writings brought him little recognition and less money while he lived. Except for a few essays and poems, he left only two longer published works, *A Week on the Concord and Merrimack Rivers* and *Walden,* the first of which he published at his own expense.

He did this fully confident that the book of his naturalist's observations during a seven-day New England boat trip would find many readers. Of the 1,000 copies originally printed, 706 ended up in the dusty attic of his parent's home in Concord. The critics of the day judged his writing to be merely a pale echo of his friend, Emerson. Only after Thoreau's death was his wisdom appreciated. His *Journal,* which most observers feel to be his finest work, was ultimately published in 1906 in fourteen volumes and contained some three million words. It was in this work that the embryonic conservationist and ecologist blossomed forth. It is here that we come to know the full Thoreau.

The private journals, not intended for publication and not published until forty-four years after his death, reveal the complex and many-faceted Thoreau and show what a rare person he was as a nature lover. As a poet-naturalist he kept meticulous notes on the various species of birds and animals that inhabited the woods around Walden Pond. His special interests ranged from the behavior of hawks to the migrations of the wild Canada goose; the recording of the regularity with which the whip-poor-wills began their evening

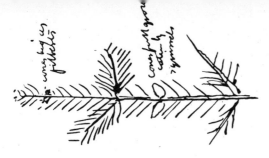

Thoreau sketch of pitch pinecones gnawed by squirrels.

songs to recording the dozen-odd items that went into making a vireo nest. As he became more serious in his observations he began to spend more time studying plants intent on becoming familiar wth every species of twig and leaf in the swamps around him. For a time he collected plant and animal specimens to send to Louis Agassiz at Harvard.

Thoreau usually preferred solitude and quiet while observing and writing about his nature findings. He repeatedly stated that the paths in the woods and his boat were his studio, but when his day's work was finished he would often seek the company of family and friends. Whenever he postponed his Concord research studies into the workings of nature and went on a journey away from his hut, he enjoyed the companionship of a congenial friend like William Ellery Channing, the Harvard theologian.

On one such hike with Channing the route was determined with the aid of Thoreau's well-used compass. The two men had started out on a trip to nearby Monadnock and went straight across fields, through swamps and over brooks until they found their compass-routed way blocked by a farmhouse barrier. The farm family, seated at their midday dinner, was astonished when the two travelers entered without knocking, went down the hall and walked out the back door without saying a word to the speechless occupants.

His classic texts, *Walden* and *A Week on the Concord and Merrimack Rivers,* testify to Thoreau's lively interest in flora and fauna but also characterize another aspect of his observations, that of the relationship of man to nature. It was difficult for him to divorce the twin interests — nature and man — for Thoreau saw them as inseparably intertwined. He believed that man derived his strength from his contact with the earth and nature, that nature is essential to man. For instance, in *Walden,* he speaks about his now famous night spent in the Concord jail for refusing to pay his poll tax in protest against the Mexican War. The incident occurred while he was living at Walden Pond. "I had gone down to the woods for other purposes," he wrote, "but wherever a man goes, men will pursue and paw him with their dirty institutions . . ." When Thoreau emerged from his night in jail, he significantly went back to nature on a huckleberry picking spree instead of seeking a soapbox to declaim against the war.

Thoreau's writing testified to his openness and love of life. In the best known paragraph he ever wrote, he said: "I went to the woods because I wished to live deliberately, to front only the essential facts of life." He believed in the power of the naked truth which he expressed over and over again in *Walden.*

Thoreau loved to be near the streams, mountains, woods and seashore — to the point that some Thoreauvians today find his panegyric endorsement of Nature a trifle excessive — yet he left a legacy for modern day Earth lovers which was best summed up in a few lines from his *Journal* . . . "I felt a positive yearning toward one bush this afternoon," he wrote. "There was a match found for me at last. I fell in love with a shrub oak."

In his approach to nature Thoreau insisted that a man begin where he is. He felt no need to leave Concord, for he sensed undiscovered continents in his own soul that needed exploring first. His going to the pond was partially an act of social rebellion — getting away from the parochial society of the small New England town. His love of nature overshadowed his utter distaste for the "lying, flattering, voting, contracting yourselves into a nutshell of civility or dilating into an atmosphere of thin and vaporous generosity."

He did not value any view of the universe "into which man and the institution of man enter very largely." In his words: "Man is but

The Thoreau cabin from the title page of *Walden*.

the place where I stand, and the prospect hence is infinite. It is not a chamber of mirrors which reflect me. When I reflect, I find that there is other than me." Thoreau thought that man was too confined to his abode. "Some rarely go outdoors, most are always at home at night, very few indeed have stayed out all night once in their lives, fewer still have gone behind the world of humanity, seen its institutions like toadstools by the wayside."

In the mid-nineteenth century he correctly predicted the unfortunate twentieth-century ecological consequences of the then still young Industrial Revolution, when he wrote in *Life Without Principles:* . . ."After reading Howitt's account of the Australian gold-diggings one evening, I had in my mind's eye, all night, numerous valleys, with their streams all cut up with foul pits, from ten to one hundred feet deep, and a half dozen feet across, as close as they can be dug, and partly filled with water — the locality to which men furiously rush to probe for their fortunes — uncertain where they shall break ground — not knowing but the gold is under the camp itself — sometimes digging one hundred and sixty feet before they strike the vein, or then missing it by a foot — turned into demons, and regardless of each other's rights, in their thirst for riches — whole valleys, for thirty miles, suddenly honeycombed by the pits of the miners, so that even hundreds are drowned in them — standing in water, and covered with mud and clay they work by night and day, dying of exposure and disease."

Thoreau decried the destruction of landscapes and wildlife wrought by the new technology. He deplored the forest waste he witnessed in the logging camps of Maine and the havoc of the shad runs caused by the dam builders on the Concord River. In this respect he was one of our first preservationists.

Thoreau was the only prominent man of letters in American history who made his living primarily from manual labor, his occupation by choice. As a combination naturalist and reformer, Thoreau believed in not accepting a life of quiet desperation but venturing out in faith, advancing "confidently in the direction of his dreams." Although he himself was not a public activist, he deeply believed that one wholly committed man could perform wonders in abolishing injustices created by his fellow man, whether it was to other men or to nature. This is what makes Thoreau so relevant today.

In Thoreau's *Walden,* which many still consider a useful handbook on how to survive in the country, contemporary ecologists can find many clues to the road to salvation and how best to use our youth as a precious resource to help make our society liveable again.

"Students should not play life, or study it merely, while the community supports them at this expensive game," he wrote, "but earnestly live it from the beginning to end. How could youths better learn to live than by at once trying the experiment of living?"

Believing that the conservation of our civilization could exist only through the reform and cultivation of the character and intellect of each man, Thoreau gave us both a method as well as a goal to serve as an alternative to the selfish goals which in his view so many men have pursued.

The Sage of Walden left a lasting legacy for modern ecologists to use as a guideline in seeking the best approaches to save our threatened environment. In a persuasive passage he wrote: "A pasture can be raucous with flowers and not make a single sound. But a man, presumably wiser than a daffodil, can beat so loudly on the eardrums that nobody hears what he is trying to say."

As he lay dying of tuberculosis Thoreau thought back about some lines that he had written in younger, healthier days and repeated them: "I wish to speak a word for

Top: Detail of map of Concord in 1852; Emerson house appears at right.

Bottom: Lithograph of Walden Pond based on a survey by Thoreau.

nature, for absolute freedom and wildness, as contrasted with freedom and culture merely civil,— to regard man as an inhabitant, or a part of and parcel of Nature, rather than a member of society . . . the walking of which I speak . . . is itself the enterprise and adventure of the day . . . I believe in the forest, and in the meadow, and in the night in which the corn grows."

In preaching the eulogy at his lifelong friend's funeral a few weeks later, Emerson prophetically said: "The country knows not yet, or in the least part, how great a son it has lost. It seems an injury that he should leave in the midst of his broken task, which none else can finish,— a kind of indignity to so noble a soul, that he should depart out of Nature before yet he has been really shown to his peers for what he is . . . His soul was made for the noblest society; he had in a short life exhausted the capabilities of this world; wherever there is knowledge, wherever there is virtue, wherever there is beauty, he will find a home."

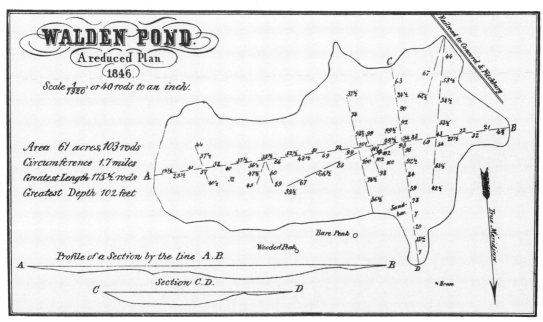

5
George Perkins Marsh

In 1847, a little known Whig congressman hailing from the Green Mountains of Vermont, wrote that "Men now begin to realize what as wandering shepherds they had before dimly suspected, that man has a right to the use, not the abuse, of the products of nature; that consumption should everywhere compensate by increased production; and that it is false economy to encroach upon a capital, the interest of which is sufficient for our lawful uses."

Thus did George Perkins Marsh point out to his fellow Americans well before the Civil War and the settlement of the West that greed and shortsightedness in the use of the land were conservation's mortal enemies. Marsh, who has been described by the urbanologist, Lewis Mumford, as "the fountainhead of the conservation movement in America," made his contribution with his classic work, *Man and Nature*, which was published in 1874.

Descended on both sides of his family from the intellectual aristocracy of old New England, Marsh was born on a farm near Woodstock, Vermont, on March 15, 1801. The son of an eminent local lawyer, John Marsh, young George took to nature in the woods around his home in the same manner as did his contemporary, Thoreau. He later wrote that, in his youth "the bubbling brook, the trees, the flowers, the wild animals were to me persons, not things."

Marsh was a frail and serious child who almost ruined his eyesight at the age of seven from too much reading. He was unable to use his eyes for long periods at a time and had to learn to listen while others read to him. At this early age, he developed interests as wide-ranging as Jefferson's.

Regaining most of his sight during his early teens, young Marsh matriculated at Dartmouth where he majored in five foreign languages: French, Spanish, Italian, German and Portuguese. He graduated with honors in 1820 and started to teach, aiming at an academic career as a college professor. But distaste for this field of endeavor, coupled with family pressures, prevailed and he commenced the study of law. He was admitted to the bar in 1825 and soon found himself a leading businessman-lawyer in Burlington, Vermont.

Marsh was not content with the life of a small town lawyer, however, and he continued to read books and scientific papers in his spare time. By the age of thirty, Marsh had already mastered almost two dozen different languages and had even found time to do several translations of European literary masters in English. Three years after settling in Burlington he married, but four years later, in 1832, his first wife died. This loss occurred within a few days after the passing of his eldest son. The twin tragedy was a crushing blow to Marsh.

Seven years later he married again, this time to Caroline Crane. She was to serve as a valuable helpmate in his intellectual pursuits and to share forty-four golden years with him, despite several illnesses brought on by her frail health. Although the marriage was childless she made a good wife for Marsh.

An interested observer of the exploitation of the surrounding countryside, he became concerned about the misuse of the land. Marsh, who was also a farmer and lumber dealer, witnessed the gouging done by his fellow woodsmen to the slopes of the Green Mountains and valleys of Vermont. He noticed precious topsoil being washed off the land and the overflowing of nearby rivers because overgrazing and overcutting had increased the run-off of water.

As its surrounding timber was depleted Burlington fell upon hard times, and it was soon forced into importing lumber from Canada to keep its economy healthy. Marsh watched this deterioration with saddened eyes seeing the necessity for wise management of resources. Although the science of ecology was still to be born, Marsh had unconsciously become one of our first pioneers in the movement seeking a balanced approach to nature.

In 1842, while he was looking for a solution to these local environmental problems, Marsh was pressured to run for Congress on a high-tariff platform under the Whig party label. He spent the next six years in the House of Representatives, where he made many useful contacts with scientists and diplomats. Marsh was opposed to slavery and, like Lincoln and Thoreau, also strongly objected to the infamous Mexican War.

While serving in Washington Marsh was deeply influenced by former President John Quincy Adams, who returned to public life as a congressman after his defeat by Andrew Jackson in the 1828 election. Adams revived Jefferson's plan for federally financed internal improvements recommending that it be expanded. He believed that land is the nation's primary resource and that it should be used for constructive purposes. To give it away to the land-grabbers was to waste the national capital. Marsh liked this idea and Adams, in turn, praised Marsh's insight into the real estate tax matter. Adams also supported the Vermonter's plea for a research-oriented Smithsonian Institution. He called Marsh's speech "one of the best ever delivered in the House," and sarcastically warned his colleagues that they "would now have to put up with what many of them had never seen before, the spectacle of a living scholar."

Unfortunately, Marsh was no great statesman. He opposed the policy of unlimited immigration, unaware that many immigrants would contribute much to the field of conservation. He was also doubtful of the ability of the American people to use the vast, undeveloped West. His Eastern Seaboard parochialism caused him to proclaim that we should learn to use our land more economically before moving too fast into an unplanned exploitation of the Western lands.

Marsh did show an interest in the eventual development of the West. In 1847 he had prepared a paper on an irrigation plan for the arid West at the request of the commissioner of agriculture. He recommended that large irrigation projects would be more practical if they were conducted on the basis of thorough hydrological surveys and on a large river-basin scale, instead of on a piecemeal approach.

He also foresaw the possibility of private monopolies endangering future water supplies and made a plea for the irrigation of the West to be accomplished "under Government supervision, from Government sources of supply."

His lawmaking career was interrupted in 1849 when President Zachary Taylor appointed Marsh as the U.S. Minister to Turkey. This appointment gave the scholar-diplomat a chance to take several side trips to the neighboring countries of the Middle East. As a self-appointed collector for the Smithsonian, he took it upon himself to send back hundreds of valuable specimens.

In 1861 Marsh, who had by now become cosmopolitan and had switched his allegiance to the Republican party, was appointed ambassador to Italy by President Lincoln. Marsh served with distinction for twenty-one years until his death in this latter post. While in Rome he accumulated an extensive and valuable collection of engravings by European masters.

But it was his other hobby of research, the impact of man on nature, pursued by this conscientious scholar over the years at his vacation home on the Italian Riviera that left his name indelibly inscribed in the history of the conservation movement. It was there that Marsh started to write a 664-page book

which was to become the foundation for the science of ecology.

Although he finished his masterwork in 1863, the publication of his book, *Man and Nature,* did not occur until 1874. (Eight years later, in 1882, a revision of this conservation classic appeared under the new title, *The Earth Modified by Human Action.*) He planned this book as a pioneer effort to "suggest the possibilities and the importance of the restoration of disturbed harmonies and the material improvement of waste and exhausted regions."

Man and Nature was written by a man of personal dignity and reserve, who combined the best qualities of the naturalist, geographer, humanist, historian and practical politician in his person. The book soon gained a worldwide reputation as a veritable encyclopedia of facts about the lands and waters of the world. The copious bibliography listed over 200 articles and works that had been published in over a dozen European countries. The contents of Marsh's book covered a wide variety of subjects: sand, soils, bogs, woods, water, canals, weather and astronomy—to name a few.

In the brilliant introduction to his work, Marsh stated the main object of his inquiry. He wanted to "indicate the character... and extent of the changes produced by human action in the physical conditions of the globe we inhabit; to point out the dangers of imprudence and the necessity of caution, in all operations which, on a large scale, interfere with the spontaneous arrangements of the organic and inorganic world."

Further along, Marsh wrote what could have been considered a preview of the ecological dilemma of the late twentieth century. "In the rudest stages of life, man depends upon spontaneous animal and vegetable growth for food and clothing, and his consumption of such products consequently diminishes the numerical abundance of the species which serve his uses. At more advanced periods he protects and propagates certain esculent vegetables and certain fowls and quadrupeds, and, at the same time, wars upon rival organisms which prey upon these objects of his care or obstruct the increase of their numbers. Hence the action of man upon the organic world tends to derange its original balances and while it reduces the number of some species, or even extirpates them altogether, it multiplies other forms of animal and vegetable life."

"The result of man's ignorant disregard of the laws of nature was deterioration of the land," Marsh wrote. "The ravages committed by man subvert the relations and destroy the balance which nature had established. . . and she avenges herself upon the intruder by letting loose her destructive energies . . . . When the forest is gone, the great reservoir of moisture stored up in its vegetable mould is evaporated. . . .The well-wooded and humid hills are turned to ridges of dry rock. . .and. . .the whole earth, unless rescued by human art from the physical degradation to which it tends, becomes an assemblage of bald mountains, or barren, turfless hills, and of swampy and malarious plains. There are parts of Asia Minor, of Northern Africa, of Greece and even of Alpine Europe, where the operation of causes set in action by man has brought the face of the earth to a desolation almost as complete as that of the moon . . . . The earth is fast becoming an unfit home for its noblest inhabitant and another era of equal human crime and human improvidence. . . would reduce it to such a condition of impoverished productiveness, of shattered surface, of climatic excess, as to threaten the deprivation, barbarism, and perhaps even extinction of the species."

Former Secretary of Interior Stewart Udall in his book *The Quiet Crisis* has observed that "To optimistic Americans, these were extravagant words. Marsh was a Jeremiah prophesying doom and it is understandable that men misled by the superabundance of a virgin continent would find his warning farfetched." Marsh was obviously a visionary ahead of his time. He did not merely concentrate, as he might have, on the mistakes of the past, but rather centered the main body of his work around the interrelationships of each natural organism and its place in the delicate makeup of earth's ecological balance.

Marsh warned us that, if man systematically destroyed any part of the pattern of nature, that the entire web of life might be transformed with unfortunate aftereffects. "Man is everywhere a disturbing agent," he wrote. "Wherever he plants his foot, the harmonies of nature are turned to discords."

He further deplored the wanton sacrifice of millions of smaller birds, "which are of no real value as food, but which . . . render a most important service by battling, in our behalf, as well as in their own, against the countless legions of humming and creeping things, with which the prolific powers of insect life would otherwise cover the earth."

Marsh also observed that if the numbers of insects were not kept in balance, that the "destruction of the mosquito, that feeds the trout that preys on the May fly that destroys the eggs that hatch the salmon that pampers

Top: A dairy farm in the East circa 1860.

Center: Logging operations in the late 19th century uncovered the slopes.

Bottom: Erosion and abandonment, the price of mismanagement.

the epicure could ultimately result in a severe shortage of edible fish."

Marsh postulated that the larger doings of man could upset nature's balance in the same manner. He stressed the impact of the draining of lakes and marshes on the wildlife habitat, the vegetation, the water table and even on the local climate. He suspected that all men's improvements, undertaken without careful thought to their effect, may well result in damage to the environment.

He also expanded on his earlier observations on the increases in such natural phenomenon as floods and soil erosion after extensive timber cutting. Marsh believed that disaster would be the inevitable result of the disturbance of the balance of nature without thought. But he also believed it possible to undo much of the harm already done by cooperating with nature.

Marsh predicted that the way out of this dilemma was for man "to become a co-worker with nature in the reconstruction of the damaged fabric which the negligence or the wantonness of former lodgers had rendered untenable. He must aid in reclothing the mountain slopes with forest and vegetable mould, thereby restoring the fountains which she provided to water them, in checking the devastating fury of torrents, and bringing back the surface drainage to its primitive narrow channels."

There was little interest or understanding on the part of scientists, government officials or laymen of the time as to how the new sciences which Marsh wrote about could be integrated for the betterment of man's condition on this planet. Marsh recognized this cultural gap and did not try to recommend an organizational solution to the problem. Rather, he announced in the preface the limits which he set for himself: "In these pages, it is my aim to stimulate, not to satisfy, curiosity," he wrote.

The scholarly author of *Man and Nature*, more than any other American of his time, served us with a clear warning that our natural resources would not go on forever. "It is certain," he said, "that a desolation, like that which has overwhelmed many once-beautiful and fertile regions of Europe, awaits an important part of the territory of the United States...unless prompt measures are taken to check the action of destructive causes already in operation."

At the end of his book, Marsh left us with a pessimistic view of things to come. "It is a legal maxim that 'the law concerneth not itself with trifles,' but in the vocabulary of nature, little and great are terms of comparison only; she knows no trifles and her laws are inflexible in dealing with an atom as with a continent or planet.... Our inability to assign definite values to these causes of the disturbance of natural arrangements is not a reason for ignoring the existence of such causes in any general view of the relations between man and nature; and we are never justified in assuming a force to be insignificant because its measure is unknown, or even because no physical effect can now be traced to its origin."

His book became a bible for the Americans who were devoted to conservation of our lands, waters and forests, and he helped to mold the thinking of such later turn-of-the-century giants of conservation as John Muir, Gifford Pinchot and John Burroughs. But it wasn't until nearly a century later that Americans of another generation began seriously to heed Marsh's warnings that we should adopt measures aimed toward the prudent management of our dwindling resources.

At least half of his book was devoted to a plea to save our woodlands as the primary endangered natural resource. Marsh observed that "the most destructive among the many causes of the physical deterioration of the earth" was the ruthless cutting down of our precious forests. He then showed how the trees served a useful purpose in preventing erosion of the land as well as serving as a natural watershed. He saw the need for prudent cutting and the immediate replanting of new trees.

Stewart Udall summed up Marsh's contributions on forestry when he noted that: "Marsh urged that American landowners reforest the woodlands in recognition of the 'duties which this age owes' to the next. He was fifty years ahead of his countrymen in this call for a national program of experimental forestry."

As the years passed after his writing of *Man and Nature*, Marsh concluded finally that only federal legislation and leadership from Washington could save our remaining forests. He had come to doubt that education alone plus enlightened self-interest would be enough to persuade the lumbering interests to change their wasteful practices.

Many of Marsh's questions concerning the long-term effects of the lack of meaningful conservation programs aimed at protecting the soil, water, climate, trees, birds and animals have still not been adequately answered to this day. His recommendations for greater research efforts into these problem areas are even more valid now than when he first proposed them. He was convinced that the emerging technologies would primarily be for the benefit of the land-raiders, who thought of their own profits first and the ultimate needs of their fellow man last.

Marsh died in his eighty-second year on July 24, 1882, far away from his native land, in Vallombrosa, Italy, near the beautiful city of Florence. He left an important legacy for others to build upon: that man was part of the cycle of nature and he should never forget it—or his own future life on this planet would be threatened.

6
John Wesley Powell

One day in 1878 a crusading, conservation-minded Secretary of the Interior in the post-Civil War Hayes' administration, named Carl Schurz, read with considerable interest a white paper report that had reached his desk. Schurz, who had come to this country as a refugee from Germany, possessed a natural conservationist bent and was favorably impressed with the paper's contents. Schurz' reaction to this plan, written by one of his subordinates, led to a drastic change in the course of the government's land policy in the West, for it set into motion an ecological revolution beyond the Mississippi whose impact is still being felt today. This historic white paper was written by a little-known federal official, Major John Wesley Powell. It was entitled: *A Report on the Lands of the Arid Regions of the United States*.

Powell's report was a plan for the utilization of the dry lands of the West. Less than twenty inches of annual rainfall is received by the land west of the ninety-eighth meridian, except for a few areas on the Pacific coast and in the mountains. This is insufficient for traditional methods of agriculture. Powell recommended that, in order to farm this arid land, it would be necessary to develop the larger streams into an irrigation system of lakes, dams and canals. This would require careful planning and financing by the federal government.

He also pointed out that the water shortage in the West meant that this scarce natural resource would have to be shared rationally by all users. He believed that a new cooperative effort and new water-rights laws would be needed by the settlers if these goals of a democratic plan for settlement were to be achieved. Powell stressed that arid Western lands were almost of no immediate value and that something should be done to prevent the cattle barons and the large land-grabbers, organized into syndicates, from gobbling up scarce water resources and the surrounding land before it was too late. He saw that the ecological balance of the West could only be maintained if development were built around a sane land and water policy.

His report also analyzed the peripheral problems of water: soil conservation, floods, erosion and hydroelectric power (which was still on the dim horizon). The late historian, Bernard De Voto, came to this conclusion regarding Powell's arid land report: "In the whole range of American experience from Jamestown on there is no book more prophetic." Part of Powell's keen vision was centered around the need to protect the small farmers and freeholders.

The cattle invasion of the plains country was already well underway when Powell submitted his report to Schurz, but Powell hoped that the harm already done could be undone. Powell's plan provided for the establishment of ranches of 2,500 acres (the size deemed necessary for the support of a family through stock-raising) and the organization of pasturage districts.

Unfortunately, Powell's land and water reform proposals like Carl Schurz' plan for protection of public forests, got short shrift from Congress. His plan conflicted with the wide open principles of the Homestead Act, and most of the powerful Western politicians, landholders and cattle barons opposed it. Powell did achieve one positive, though indirect result by giving a boost to science.

Although his major recommendations were not heeded, the U. S. Geological Survey was established on his suggestion for one responsible agency to conduct basic research into the makeup of the earth itself and to carry out a national mapping program so that meaningful resource planning could take place. The creation of this new agency of government marked a beginning of a tax-supported scientific base in the federal system. Powell felt that the facts derived from such basic research could benefit the entire nation.

The birth of the American Association for the Advancement of Science in 1873 also owed much to the pioneering work of Powell, who saw science as capable of the renewal and enlargement of all natural resources.

Who was this man, Powell, who tried to bring some sanity into the settlement of the West? He was born in Mt. Morris, New York, on March 23, 1834, the son of an English Methodist minister who had emigrated to America just four years earlier. As a young man, Powell thought he might become a minister or public speaker like his father but, after he became attracted to botany, he joined the Illinois State Natural History Society in 1854 and was soon elected its secretary. He roamed all over the state observing and collecting natural specimens. Powell also made exploratory boat trips on the Ohio and Mississippi rivers while still in his early twenties.

With the onset of the Civil War, he enlisted on the Union side, was commissioned and was soon in command of a battery of artillery. He was wounded at the bloody Battle of Shiloh, where he lost his right arm at the elbow. He returned to duty and fought through the rest of the war and wound up a major.

After leaving the army in 1865 Powell became a professor of geology at Illinois Wesleyan College where he, his wife and young daughter settled down to the routine of a quiet campus life. But two years later he was on the move again, when he conducted the first of a series of geological field trips, consisting of a group of students and amateur naturalists, to the plains and mountains of Colorado. General Ulysses S. Grant, his old commanding officer, furnished him with troops to protect his expedition from marauding Indians.

While on a second scientific field trip in 1865 Powell conceived a daring scheme, an idea not even the Indians had ever attempted:

37

Powell's Colorado expedition navigated some 900 miles of unknown waters.

descending the Colorado and Green rivers in boats.

On May 24, 1869, the party of ten adventurers in four specially designed boats put off from Green River City in Utah. Three months later, the Powell party emerged from Black Canyon, east of Las Vegas. The expedition had navigated some 900 miles of treacherous rapids and unknown waters. They had suffered from twin hardships of summer heat and a shortage of food due to the loss of supplies. The only lives lost were those of three men who had given up near the end and had climbed out of the canyon only to be waylaid and killed by Indians.

The immediate geological results of his Grand Canyon venture were slight, but the psychological and political impact was much greater. Since his journey had been conducted under the auspices of the Smithsonian Institution, Powell's standing benefited from the Smithsonian's prestige. The publicity of the journey which was widely reported in the press also helped Powell to obtain political support for his projects. In 1871 Powell returned for further exploration of the Colorado plateau.

Six years after the conclusion of his first journey, Powell published his report, *Explorations of the Colorado River of the West* (1875). In this report, he made a unique contribution, which marked him as a geologist of note, by calling attention to the fact that the Grand Canyon was created by the erosive action of the river on the sandstone rocks over eons of time. He also postulated that the rocks were constantly being elevated while the river preserved its old level.

This one-armed geologist-explorer was also interested in the Indian tribes scattered throughout the West. While most of the nation was seemingly intent on erasing all Indians, Powell was their friend. To help carry out a scientific investigation of these tribes and their predecessors on the American continent, Powell conceived the need for the establishment of a research bureau on Indian culture. In 1879, he was picked to head the new Bureau of Ethnology at the Smithsonian Institution and many of his ideas on ethnology were found in its later reports.

Since he headed the effort toward reorganization of rival surveys in the West into the United States Geological Survey, it was natural that Powell receive the appointment as its first regular director in 1880. He remained at that post for fourteen years until his retirement in 1894, leaving a successful record as its initial administrator. While holding down this post Powell encouraged his subordinates to attain personal distinction on their own.

Powell cut a rough and striking figure in public, wearing an unruly beard and a wild hairdo. He was a hearty, magnetic person and

Boat used by Powell to descend the Colorado River.

he ran the Survey like a military commander in battle. A strong disciplinarian, he abhorred sloppiness on the job. Powell, an ambitious man, was described as "electric with energy and ideas."

Although he had not been successful in his first attempt during the eighteen seventies to win his argument with the Congress for the establishment of a sane Western land and water policy, the political climate in Washington slowly changed in the wake of the severe droughts of the eighties and nineties. The validity of his earlier warnings was finally driven home. In 1888 Congress, in the face of a crisis, put him in charge of an Irrigation Survey which was empowered to carry out part of Powell's plan, selecting sites for reservoirs and irrigation projects.

Powell believed this could be accomplished in six or seven years and that the cost would be as many millions of dollars. He spoke to the constitutional conventions of Montana and North Dakota the following year, preaching the necessity of irrigation. He believed that each landowner should have a corresponding water right, and that county lines should follow the drainage pattern of the land, so that political units would not be working against regional planning. His ideas on water rights were later embodied in the "Wyoming doctrine" which was adopted by most of the Western states.

The calendar and his enemies in Congress ganged up on him, however. Powell believed that with one more year's work he might have been ready to transform some of his plans for the Western river valleys into an operational phase. Unfortunately, some members of the anti-Powell forces in Congress would not tolerate his "advanced" ideas any longer and in 1890 they quietly killed his Irrigation Survey by cutting off funds. His General Plan was put on the shelf and he went back to his scientific work at the Geological Survey.

In 1894 Powell retired from the Survey, partly for reasons of health and partly because of the frustrations which arose over the antagonism to his conservation and irrigation projects. He decided to spend the rest of his life writing about anthropology in his old office at the Smithsonian on the Mall.

Yet Powell's basic ideas had taken root, and they flowered in 1902 when Congress finally passed the Reclamation Act. That act did not contain everything that Powell had fought for, but it did mark a beginning. Ironically, this event occurred in the year of his death. He died from a cerebral hemorrhage while he was vacationing in Haven, Maine, during his sixty-eighth year.

Thirty years later, his unheeded warnings on the misuse of the land came to tragic fruition when sand and dust storms swept over the semi-arid high plains creating the "Dust Bowl" of the '30s. Land which had been overgrazed or unwisely put to the plow had lost its protective sod. A long, dry spell aggravated conditions and powder-like soil yielded to high winds bringing disaster.

Wallace Stegner, in his book, *Beyond the Hundredth Meridian,* gave this appropriate tribute, when he wrote of the man: "Powell was one of those powerfully original and prophetic minds which, like certain streams in a limestone country, sink out of sight for a time to reappear farther on.... He tried to shape legal and political and social institutions so that they would accord with the necessities of the West. He tried to conserve the West's natural wealth so that it could play to the full its potential part in the future of the United States. He tried to dissipate illusions about the West, to sweep mirage away. He was a great man and a prophet. Long ago he accomplished great things and now we are beginning to understand him...even out West."

7
Guy Bradley

On July 8, 1905, a series of distant shots crackled faintly in the still summer air across the calm waters of Florida Bay off the tip of East Cape Sable in southern Florida. Guy Bradley, thirty-five, one of four game wardens in the area hired by the Audubon Society to stop the slaughter of the American and snowy egrets for their beautiful white plumes, reached for his nickel-plated .32 caliber revolver. He gave his wife, Fronie, a quick kiss and his two small sons a pat on the head before jumping into his rowboat to take off toward the sound of the shooting.

Near Oyster Keys he saw a small schooner anchored, and he recognized it immediately as belonging to Walter Smith, a man who had threatened to kill Bradley if he ever interfered with his egret poaching in the rookery. As Bradley drew near the law-breaking plume-hunter's boat, Smith fired a shot into the air as a warning to his two sons, who were ashore in a skiff picking up dead birds from a rookery on the keys. The skiff returned with Smith's two sons, Tom and Dan, and Bradley's little boat pulled alongside the bearded seaman's schooner, *Key West,* before it could get underway.

Smith, who had been arrested once before by Bradley for shooting in the rookery, asked the game warden what he wanted.

"I want your son, Tom, the one with the birds," said Bradley calmly, "he's under arrest."

"You ain't got no warrant," Smith shouted back from the schooner's deck while holding a Winchester rifle in his hand. "I'll be damned if you'll get him!"

After some further brief talk, two shots rang out almost simultaneously. Bradley's .32 caliber bullet lodged in the schooner's mast, but the bullet from Smith's heavy rifle entered the left side of Bradley's neck, ranging downward through his body and killing him instantly. The little boat bearing the crumpled body slowly drifted with the southeast breeze to Sawfish Hole, near East Cape Sable, where the Roberts boys found it the next day, being attracted to the spot by a flock of vultures wheeling in the air overhead.

After the shooting, Smith upped his anchor and sailed back to Key West. When word of the finding of the body reached the mainland, newspapers in Key West, Palm Beach, Miami and New York featured the story. Smith was put in jail at Key West in lieu of a $5,000 bond, pending action by a local grand jury. In Key West the papers headlined: "Indignant Neighbors Burn Smith's House — Flamingo People Incensed Over the Killing of Guy Bradley." A Negro and his wife, tenants in Smith's house, were ordered to move out by the local residents who soon applied a torch to its contents as a protest to the brutal murder of Bradley.

Five months later, after the grand jury deemed the prosecution had failed to "present sufficient evidence to bring the accused to trial and failed to present a true bill," Smith was released and in a sense was acquitted. Smith testified that Bradley had fired first and that he fired back in self-defense.

The prominent American ornithologist, Frank Chapman, had uncannily predicted a year earlier in the *New York Sun* that "That man Bradley is going to be killed sometime. He has been shot at more than once, and some day they are going to get him."

As a token of its feelings concerning Bradley's sacrificing his life to save a species of wild fowl from extinction, the National Audubon Society bought his widow, Fronie, the cottage of her choice in Key West. Three years later, another Audubon warden, Columbus G. McLeod, was killed by plume hunters near Charleston, South Carolina, while on a mission similar to Bradley's. The deaths of these two men helped to arouse public opinion over the need to protect these rare and

beautiful birds from extinction.

While still a baby, Bradley had come to Florida with his parents from Chicago in 1870. They emigrated southward shortly after the disastrous fire had hit the Windy City. His father finally settled in Lantana, near Miami, and for a time was superintendent of schools in Dade County. In those days — 1880 — the population of that county was only 257 as compared with the million-plus people who live there today. Bradley and his older brother, Louis, loved to fish and hunt as well as to play the fiddle in a string band which performed at dances and parties in Palm Beach.

The elder Bradley, motivated by the lure of free land at the southwest tip of the Florida mainland, moved his family to the shores of Florida Bay, where he and his boys grew potatoes, tomatoes and eggplants in the rich hammock land bordering the waterfront. Those settlers who were less industrious lived by fishing and hunting white egrets for their plumes.

By 1900 the tiny settlement of Flamingo on Cape Sable boasted a one-room schoolhouse and a post office. Edwin Bradley was the postmaster and his elder son, Louis, sailed the mailboat to Key West. One day a strange schooner put into Flamingo with two brothers named Vickers and their sister, Fronie, on board. Guy Bradley soon fell in love with the girl and they were married. They became proud parents of two sons, Ellis and Morrell.

At the turn of the century the beautiful plumed birds — the American and snowy egrets and the roseate spoonbill — had almost become extinct in south Florida because of the tremendous rate of slaughter being carried on by illegal hunters who catered to the demands of the New York millinery trade.

In 1892 alone, one "feather merchant" from Jacksonville shipped 130,000 "scalps"

The Everglades region was an unspoiled sub-tropical wilderness until hunters found that bird plumes were worth their weight in gold.

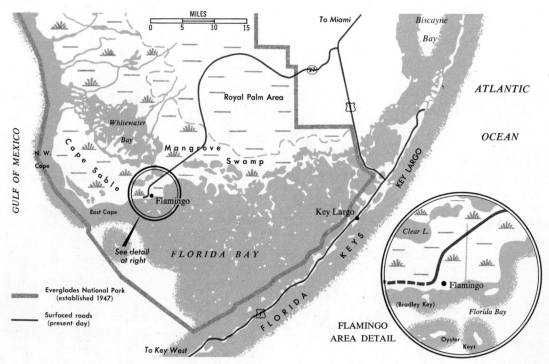

(egret skins with feathers attached) to New York. Hunters usually got $1.25 per skin and one firm sold $200,000 worth of egret plumes to eager customers in a single year. In the pursuit of quick profits, White and Indian hunters often left thousands of young egrets to starve in their nests, since the parent birds only sprouted plumes during the nesting season. At this time Florida had no effective law to prevent such slaughter of the egrets.

Shocked finally by the revelations of the Audubon Societies and the American Ornithologists' Union, the Florida legislature in Tallahassee passed several laws in 1901 protecting non-game birds such as egrets. These laws are still on the books.

Unfortunately the state did not allocate sufficient funds for wardens. There was no Florida Game and Fresh Water Fish Commission to administer the new laws, but the governor did appoint one game warden for each county. In an area like Monroe County, which included vast stretches of mangrove and everglade wilderness on the mainland as well as the long strip of Florida keys, obviously could not be served effectively by one lone game warden. For this reason the National Audubon Society, at its own expense, raised money to employ four wardens to help protect the endangered egrets. Bradley, then thirty-two, was the first to be hired. He knew what he was getting into — the egret plume hunters were dangerous characters.

With the slaying of Bradley and McLeod national attention was brought to the protection effort. The Audubon Society then struck at the heart of the traffic in bird "scalps," the millinery trade in New York. The attack began in Albany where they saw to it that the "Audubon Plumage Bill" was introduced in the state legislature. T. Gilbert Pearson, the president of the society, spearheaded the fight for passage of the bill which outlawed the commercial use of wild bird

Preservation of wildlife in America did not begin until the early 1900s.

feathers in New York State. The bill was passed two years after McLeod's death (in 1910) and signed into law by Governor Charles Evans Hughes. Traffic in bird plumage was soon stopped everywhere.

As tribute to the first American martyr to the cause of conservation, a wayside exhibit exists today in the Everglades National Park as a memorial to a man who gave his life to save a rare species of beautiful tropical birds from extinction. The wild Cape Sable country where Bradley roamed as a boy and as a young game warden is today preserved forever as part of the Everglades National Park.

Today the Audubon Society sponsors Everglades wildlife tours which visit a network of Audubon bird sanctuaries within the park. In a way these sanctuaries are the legacy of the noble work of Bradley and McLeod. It was the martyrdom of these men that advanced the legislation needed to save many species from extinction.

On the simple stone marker, located at the site of Bradley's grave on Cape Sable, the following commemoration is emblazoned on a brass plaque:

"Guy M. Bradley
1870-1905
Faithful Unto Death
As Game Warden of Monroe County
He Gave His Life for the Cause to
Which He was Pledged."

Some conservationists fear that endangered birds such as the whooping crane, the California condor, the ivory-billed woodpecker, the pelican and the American bald eagle are already too far reduced in numbers and are destined to join the passenger pigeon in extinction. The bird and animal protection effort continues, however, as man becomes increasingly involved in working for survival of species, his own included.

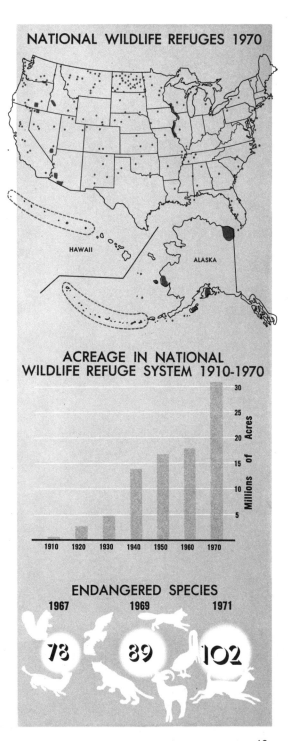

8
John Muir

Naturalist, conservationist, explorer. These words describe a canny Scottish emigré to America who came to this country as an eleven-year-old boy and stayed on to become a drumbeater for the setting aside of large tracts of virgin American forests as national parks for later generations to enjoy. John Muir was that man.

John Muir was born in Dunbar, Scotland, on April 21, 1838. He was the oldest son in a family of eight children which included five daughters. His stern father often exerted severe corporal punishment on John if he didn't daily memorize assigned verses from the King James version of the Bible. In 1849 his parents emigrated to America where they settled on a Wisconsin farm. There he learned to be a railsplitter and to plow, hoe, mow and cut grain.

Self-taught, John was prevented from studying at night by his father's requirement that he get up at dawn. Muir then invented an ingenious wooden clock, which he attached to his bed as a crude alarm or "early-rising machine" as he called it, so that he could awaken in time to study before embarking on his early morning farm chores. When he exhibited this and other mechanical inventions at the Wisconsin State Fair in 1860, he came to the favorable attention of the University of Wisconsin authorities.

Muir later enrolled at the University of Wisconsin, one of our leading land-grant, conservation-minded colleges. He majored in chemistry and geology but did not complete his degree partly because of his dislike of required courses and the lack of freedom to choose his own. At the close of his short college career, just before leaving in 1863, Muir became passionately absorbed in botany —a field that would influence his later career.

After leaving Madison, Wisconsin, he proceeded to travel extensively on foot through the Midwest and Canada. Muir took occasional jobs and continued his interest in mechanical inventions, until a machine accident, suffered while working in an Indianapolis factory in 1867, made him fear the permanent loss of an eye. From then on he decided to "bid adieu to mechanical inventions" and to devote the rest of his life to "studying the invention of God" in nature.

He then commenced a long trek southward to the Gulf of Mexico. This journey, including his observations of the flora and fauna en route, was recorded and published many years later in the form of a daily journal called: *A Thousand Mile Walk to the Gulf*.

From there he wandered westward to California, arriving in the Golden State in 1868. He spent the next six years in a refreshing and enthusiastic study of the glaciers and forests of the Sierra Nevadas. The spectacular Yosemite Valley was his home camp during these important years. Later he moved on to the American Northwest and Alaska. While in the far North he explored the upper stretches of the Mackenzie and Yukon rivers. During his travels he kept copious journals on all of his observations and illustrated them with sketches.

He finally tired of this exhausting, nomadic existence, and, after marrying Louie Wanda Strentzel in April 1880, he decided to settle down for a while. She was the daughter of a Polish emigré who had become a California horticulturist. In 1881 Muir rented some land from his father-in-law and later bought the Strentzel fruit ranch. He soon learned the art of fruit growing and prospered at this enterprise for the next ten years. Then in 1891 he sold part of the ranch and leased the rest so that he, his wife and two daughters could devote more time to travel and his writing.

He spent many of his remaining years in round-the-world jaunts, going to such far off places as Australia, South America, Africa and Siberia to study trees, shrubs, forests and

44

Below: The famous Yosemite Valley was sculptured by glaciers. John Muir was instrumental in making the area a national park.

glaciers. The latter two pursuits became his principal interests as he grew older.

Muir was the first to demonstrate the true glacial erosion origins of the Yosemite Valley. When in the early 1870s Muir declared unequivocally that the Yosemite Valley was the result of glacial action, the eminent geologist, Josiah Whitney, tried to discredit him with the twin epithets: He (Muir) is "an ignoramus . . . a mere sheepherder!"

"A more absurd theory was never advanced," Whitney sputtered. He insisted, rather, that the Yosemite Valley had been formed during a single "grand cataclysm" when the earth's crust collapsed in the "wreck of matter and the crush of the worlds."

Muir, however, was equally stubborn and certain that the magnificent California valley, which many consider to be our most awe-inspiring national park, had been carved in a much less spectacular fashion. Muir's theory was that such outstanding prominences as Half Dome and El Capitan were carved by a slow-moving river of ice over a million years ago.

Neither could have known at the time that Whitney's hypothesis would soon be severely tested. By 1872 the narrow gash in the Sierra Nevada range with rock walls 3,000 to 4,000 feet high had been known to the white man for only seventeen years, although Indians had lived there far longer. Horses could descend its steep mountain trails, but the first wagon road was still several years away from completion.

Then at 2:30 a.m. on Tuesday, March 26, 1872, during a moonlit, wintry night, the first shock of an earthquake jolted the few residents awake. This Yosemite or Inyo earthquake was later believed to be of equal intensity to the more famous San Francisco earthquake that occurred some thirty-four years later in 1906.

If Josiah Whitney's theory of massive block faulting was correct, all the people in the valley might soon perish. Their immediate dilemma was whether to stay in the valley or to flee. In less than one minute on that chilly night, large boulders began to shear off Yosemite's towering walls. Eagle Rock on the south wall gave way with a tremendous "rock-roar" greater than "all the thunder of all the storms I had ever heard," so Muir later recorded in his eyewitness guidebook, *The Yosemite*. "The whole earth, like a living creature, seemed to have at last found a voice and to be calling to her sister planets," he wrote poetically about the incident.

Eagle Rock ruptured immediately into thousands of smaller boulders which showered to the valley floor in a great arc. Muir was asleep in his valley cabin when the tremors began. Awakening, he ran outside "both glad

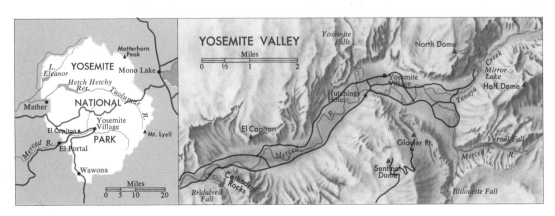

and frightened" shouting, "A noble earthquake." He felt certain that he "was going to learn something."

The shocks were violent and so close together that Muir balanced on the swaying ground "as if on the deck of a ship among waves . . . it seemed impossible that the high cliffs of the valley would escape being shattered."

But fortunately, the young naturalist's curiosity conquered fear, and Muir ran up the valley, clambering over rocks still "slowly settling into their places, chafing, grating against one another, groaning and whispering . . . a cloud of dust particles, lighted by the moon, floated out across the whole breadth of the valley. . . . And the air was filled with the odor of crushed Douglas spruces from a grove that had been mowed down and mashed like weeds." He went to the bank of the Merced River "to see in what direction it was flowing" and was relieved "to find that *down* the valley was still down."

Twenty-seven people were killed and sixty injured in the then sparsely settled Yosemite Valley. The village of Lone Pine was destroyed as the earth moved laterally as far as twenty feet and vertically as much as twenty-three feet. In the gray dawn, the few, year-round, thoroughly shaken Yosemite residents gathered at the local Hutchings Hotel, wondering if they should flee or stay. Was this a new episode in the valley's cataclysmic creation? At the moment, Whitney's theory about the formation of the valley seemed correct. Whitney had been California's official state geologist for twelve years and had been widely acclaimed for his clear and logical explanation of the origin of the valley which he described in the official *Yosemite Guidebook*, which he had written.

Mount Whitney, the highest peak in the United States, bore his name in honor of his geological survey of the state conducted be-

Grasshopper sketch by John Muir on a letter to a close friend.

tween 1864 and 1870. He had also served on the faculty of Iowa State University and later was to become a professor of geology at Harvard.

Muir, on the other hand, was merely a forest guide and a sawmill operator—to some, just a sheepherder who did odd jobs at the hotel. Although he talked much about nature and science, his main writings at the time had consisted of letters to his family and friends and notes in his journal.

During the urgent discussion that was held in the Hutchings Hotel, no one present knew that in the course of time John Muir's name would far exceed Whitney's and that future geologists would find that Muir's analysis was closer to the mark than Whitney's. Muir had erred in estimating the relative amounts of glaciation and stream erosion in the range, since ice had never covered the Sierras totally from foot to summit. Rather, three invasions of ice a million years ago had been responsible for at least a third of Yosemite's depth and the rest came from the slow, eon-long carvings of the Merced River.

When Muir arrived at the hotel another series of shocks "made the cliffs and domes tremble like jelly." Pine needles "flashed and quivered in the sunlight" while the "branches waved up and down although the trunks stood rigid." The men in the hotel were silent. "The solemnity on the faces was sublime," Muir observed in his book, *The Yosemite*.

Whitney's "wild tumble-down-and-engulfment hypothesis might soon be proved," he told them. "These underground rumblings and shakings might be the forerunners of another Yosemite-making cataclysm, which would perhaps double the depth of the valley by swallowing the floor, leaving the ends of the roads and trails dangling three or four thousand feet in the air."

Then came a third series of shocks. Birds flew frantically from the trees. Seeing his "well meant joke" had worried the men, Muir tried to calm them and belittle the earthquake. "Mother Earth is trotting us on her knee to amuse us and make us good," he said.

Muir, of course, did not flee the valley. He pointed out that the granite walls were the most solid masonry on earth, far less "likely to collapse and sink than the sedimentary lowlands," which everyone else there felt offered more safety. "In any case," he said calmly, all men "sometime would have to die, and so grand a burial was not to be slighted."

One doubting resident, however, decided that Whitney had indeed been correct and, convinced that the valley was sinking, shouted, "I'm going to the lowlands until the fate of the poor trembling Yosemite is settled!" After locking the door of his cabin, he gave his key to Muir to watch his place, just in case the sheepherder was, after all, right. Time proved Muir correct and Whitney's thesis of cataclysmic changes was soon discarded.

In 1889 Muir took Robert U. Johnson, an editor of the *Century* magazine, to Yosemite Valley and showed him another type of devastation wrought by uncountable hordes of sheep. At Johnson's request, Muir started a propaganda campaign to establish a permanent Yosemite national park. His subsequent series of brilliant articles in *Century* on the value of the Western forests and their spectacular scenery had a profound educational effect on the public mind. The result was that many public-spirited men in the country rallied around to support the ideas of Muir and Johnson, and they lobbied Congress until the Yosemite National Park was established in 1890.

A year later, Congress passed another act empowering the president to create forest preserves. Back in 1876 Muir had proposed that a national commission be appointed to look into the fearful wastage of our forests,

Muir, President Roosevelt
and others at Mariposa Grove.

to survey the forest lands and to recommend conservation measures.

In 1896, twenty years later, a presidential commission was finally appointed and chaired by Charles Sargent. He invited Muir to accompany his party on an investigative tour of the forests. The following year, on the basis of this commission's recommendations, President Cleveland created thirteen national forest preserves comprising 21 million acres.

The predatory commercial interests tried to nullify these preserves through congressional action, and they even succeeded in restoring to the public domain all the forest reservations, except those in California, until March 1, 1898. They felt victory was theirs.

The battle was now joined, in Muir's phrase "between landscape righteousness and the devil." Here, he found his greatest opportunity for rendering a notable public service. He wrote two articles, one published in *Harper's Weekly* in June 1897 entitled "Forest Reservations and National Parks" and the other in *Atlantic Monthly* in August 1897 called "The American Forests." These two articles turned public sentiment in favor of ending the plunder of the forests. Muir was by now the acknowledged leader of the conservation movement in the United States.

"His style rose to the impassioned oratory of the Hebrew prophets arraigning the wicked in high places and preaching the sacred duty of using the country we live in that we may not leave it ravished by greed and ignorance, but may pass it on to future generations undiminished in richness and beauty," observed one contemporary admirer.

His invective won the ear of Congress, and in 1898 the enemies of forest reservations failed to extend the earlier acts of annulment when Representative John Lacey, the Chairman of the Public Lands Committee in the House of Representatives, insisted that Muir's judgment was better than the opposition.

This "bearded zealot who preached a mountain gospel with a John the Baptist fervor," as he was characterized by Stewart Udall, set out to form a private organization of mountaineers and conservationists to carry on his fight to protect the wilderness. Out of this effort came the Sierra Club, a crusading organization "to explore, enjoy and protect the nation's scenic resources."

The formation of this club, one of the two leading conservationist organizations of the present day, caused Muir to enter into a new phase of his career. No longer was he just a writer-naturalist. Now he was an organizer and a publicist leading the fight against the lumbermen and stockmen who wished to take over the Yosemite country.

In the last decade of the nineteenth century Muir carried on a national campaign concerning the human wastage of the natural forest resources. His writings and lectures helped influence the general trend toward conservation in Congress and the White House.

In the spring of 1903 he went camping with President Roosevelt in Yosemite Park and he used the opportunity to expound his views with his sympathetic listener. During Roosevelt's administration, 148 million forest acres were set aside, sixteen national monuments were established and our national parks were doubled in number—due in great part to Muir's untiring efforts.

Muir, a lithe man of medium height and spare of frame, possessed extraordinary powers of physical endurance. His auburn hair and blue eyes added to his engaging personality. His quick repartee and interesting conversation fascinated his listeners.

He suffered one noticeable defeat during the last six years of his life, when he made an unsuccessful attempt to preserve the beautiful Hetch-Hetchy Valley in Yosemite. A combination of commercial interests, the Army Corps of Engineers and political intrigue won out, and a man-made dam turned the valley

Muir and Roosevelt on the trail below Half Dome in Yosemite.

into a water reservoir for San Francisco.

One good result came from this failure, however. Public sentiment became consolidated against other such raids on scenic wonders like the Hetch-Hetchy Valley.

Muir died in Los Angeles on December 24, 1914, at the age of seventy-six, having completed a life full of worthwhile contributions for the welfare of his fellow man. A beautiful redwood forest, located on the peninsula just north of the Golden Gate on San Francisco Bay, had been named Muir Woods in his honor in 1908. He loved the giant sequoias and pines of California and cherished this recognition.

Today, while time to save our land and forests is growing short, we should heed the words of John Muir:

Everyone needs beauty as well as bread
Places to play in and pray in
Where nature may heal and cheer and give
   strength
To body and soul.

9
John Burroughs

"He was truly a priest of nature," said one of his close friends. At the time of his death in 1921, John Burroughs, the originator of the nature essay, had acquired a greater personal following in the United States than had any other contemporary author. This popularizer of nature's everyday doings played a vital role as a publicist for the cause of conservation, following in the footsteps of Thoreau and Emerson.

Burroughs served as a dispenser of popular information concerning nature study and helped to extend this knowledge both within and outside the schools of the day through his writings. Besides his major contribution of spreading the news about nature and conservation to the American public, Burroughs also performed a secondary role in stimulating a popular interest in science in general.

The future nature essayist and conservationist was born on a farm situated in the lower Catskill Mountains near Roxbury, New York, on April 3, 1837.

The seventh born in a family of ten children, Burroughs was descended from a long line of New England settlers dating back to the early seventeenth century. From his paternal ancestry he felt that he derived a love of peace, solitude and high intellect. From his maternal side he believed he acquired his love of nature and idealism. These qualities dominated the personality of the future naturalist-publicist as he matured.

During his boyhood he loved to roam the hills surrounding his home. Young Burroughs delighted in spending hours watching birds and exploring the nearby woods. In 1854, when he was in his late teens and searching for a career, he taught for a short time in a small country schoolhouse in Ulster County, but then he left to attend the Ashland College Institute. In 1856 he entered Cooperstown Seminary in the town where baseball was born. While there, he read St. Pierre's *Studies of Nature* and Emerson's *Essays* and soon found himself in "a sort of ecstasy."

At age nineteen he left to teach school once again, this time in the village of Polo, Illinois. A year later he met and married Ursula North, who was thirteen years his senior.

When he was twenty-three he wrote an unsigned Emersonian-style essay, "Expression," for the *Atlantic Monthly*, which was published in the November 1860 issue as Emerson's own work. Then a New York periodical accepted for publication his essay called "From the Back Country." Burroughs was to have many of his essays appear in *Atlantic Monthly* over the years. He later described the *Atlantic* as his "university."

During those early years he engaged in various pursuits—journalism, farming, fruit raising. None, including teaching, seemed to satisfy. In 1863 he became absorbed with wildflowers, partly due to the influence of his botanist friend, William Eddy. He soon found a second interest when he chanced upon Audubon's *Birds of America* while browsing through the U.S. Military Academy's library at West Point. From that moment on he was a bird fancier.

It was at this time that Burroughs accepted a clerk's position at the Currency Bureau of the United States Treasury and moved to Washington. While living in the nation's capital he formed the most important friendship of his life with the poet Walt Whitman. The bearded Sage of Camden was forty-four at the time, almost twice the age of Burroughs. They took long walks together and conducted equally lengthy conversations about nature and the world.

Burroughs often accompanied Whitman on his army hospital tours, where the latter served as a male nurse to the wounded Union soldiers from the battlefronts. Burroughs commented later about their deep relationship.

"I loved him as I never loved any man," he said. "I owe him more than any man in the world. He gave me things to think of, he taught me generosity, breadth and an all-embracing charity." Significantly, when Burroughs published his first book, it was *Notes on Walt Whitman as a Poet and Person* (1867).

In the spring of 1865 he published his first nature essay, "With the Birds," in the *Atlantic Monthly*. This article became the opening chapter in his first book on nature called *In The Hemlocks*. This book and those which followed helped Burroughs escape from the clerical confines of the U. S. Treasury. The accuracy of his observations of birds and other creatures of nature led contemporary observers to conclude that he was superior to Thoreau in his poetic feeling and literary expression.

By 1866, however, his interest in birds had waned temporarily and he was not to publish his first bird book until five years later. When *Wake Robin* appeared in 1871 it received very favorable reviews from the critics. William Dean Howells wrote of Burroughs' latest work: "The dusk and cool and quiet of the forest seem to wrap the reader."

After nine years of bureaucratic grind as a government clerk Burroughs left Washington. In 1873 he purchased a nine-acre farm located near Esopus, New York. This picturesque

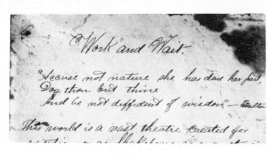
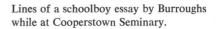

Lines of a schoolboy essay by Burroughs while at Cooperstown Seminary.

farm, which he named "Riverby," was situated on the west shore of the Hudson River about eighty miles north of New York City.

Burroughs built himself a writer's retreat, which he called "Slabsides," about a mile up the hillside from his home. During summers he went back to his old family home near Roxbury where he used the hay barn as a literary workshop. But it was at Slabsides where he did most of his contemplating and writing. It was here in his later years, after he achieved status as a leading prophet and naturalist, that many friends and admirers came to visit the "Sage of Slabsides." They reported on the erect figure with the white-flowing, rustic beard, walking around his country place with an air of repose, at home with nature and the world around him.

Burroughs usually wore simple clothes and a rakish, Western-style hat in public. A quiet, shy man, the naturalist liked to hole up in his cabin "office," as he liked to call his twelve-by-twelve-foot retreat overlooking the Hudson. Slabsides' name came from the fact that the outside walls were veneered with walnut bark that he had salvaged from some trees which were removed from his land by a local furniture company. Slabsides had only two small windows and a large stone fireplace, which took up half of a side wall.

Free at last from having to make a living in a mundane way, Burroughs traveled widely to Bermuda, Jamaica, Hawaii, Canada, the Maine Woods, Alaska and Europe. He camped in the Yosemite and Yellowstone parks with his friends, John Muir and Teddy Roosevelt, and even wrote a book about the latter *Camping and Tramping with Roosevelt* (1907). As a sage and prophet, he counseled T. R., E. H. Harriman, the railroad magnate, and other nature enthusiasts on the need for conservation.

He became closely associated with President Roosevelt in a joint campaign against the "Nature Fakers," writers who embroidered the facts of natural history. This period of his life was highlighted by his article, "Real and Sham Natural History" which appeared in a 1903 issue of *Atlantic Monthly*.

The first phase of Burroughs' literary career had ended in 1877 with his *Birds and Poets*. Then began a second period, featuring *Locusts and Wild Honey* (1879) and ending in *Leaf and Tendril* (1908). His main passion during this period was straight seeing and straight thinking about nature. He made good use of his senses in recording his keen observations of the life around him, adopting the methods of science.

A third, overlapping period began roughly in 1900 with the publication of *The Light of Day,* which reflected his enthusiasm for science and the influence of Darwin. In this volume, he endorsed Victorian science and attacked the authority of revealed religion. He stated that reason alone could give one the true interpretation of the universe. At this stage, Burroughs believed that science had bypassed both religion and literature and was in the "heat of its forenoon work," helping man explain his role and destiny.

Yet, twelve years later, he felt that science had reached its meridian and in his *The Summit of the Years* (1914), he conceded the achievements of natural science, but criticized man-made civilization as "an engine running without a headlight" that may ultimately draw man to his own destruction.

In his mature years, Burroughs believed that the salvation of society depended on vision and "the intuitive perception of the great fundamental truths of the inner spiritual world." So, he had come full circle, saying that we must look not to those who are intent on exploring life in terms of physics and chemistry, but to "the great teachers and prophets and poets and mystics," such as Walt Whitman.

After the conclusion of World War I he published his last four books, which encompassed a melancholy attempt to reconcile scientific reason with a belief in intuitive philosophy as the best route for man's salvation. He felt that "the heart often knows what the head does not. Hedge or qualify as we will, man is part of nature.... The animals live by instinct and we largely by emotions, but it is reason that has placed man at the head of the animal kingdom."

The lover of earth, the prophet of nature, "the Seer of Slabsides," a humanist who revered his fellow man—all of these descriptions applied equally well to John Burroughs. On his many excursions into the primeval wilderness of an earlier America, this "watcher of the woods" left an indelible imprint for us to follow. He discovered no new trails, no new animals or birds, but he saw the fruits of nature in a new light.

Burroughs wrote his own informal self-analysis in a letter to Dallas Sharp, the editor of *Atlantic Monthly*, whom he took to task for disparaging his idol Thoreau—the real father of American conservation.

"Thoreau is nearer the stars than I am," Burroughs wrote. "I may be more human, but he is certainly more divine. His moral and ethical value I think is much greater and he has a heroic quality that I cannot approach."

Perhaps no more generous word was ever spoken about one great writer by his nearest rival as America's outstanding naturalist-writer than Burroughs' tribute to the author of *Walden*.

From 1873, when he moved to Riverby, he was able to average writing one book every two years for the rest of his life. During that fifty-year period, most of his works were on nature subjects.

In his last years, he made friends with Thomas Edison and Henry Ford. Burroughs also received several honorary degrees from Yale, the University of Georgia and other colleges.

In his last essay, "Facing the Mystery," Burroughs recanted his earlier emotional enthusiasm for Whitman, the poet of immortality, and accepted the prospect of his own personal extinction.

John Burroughs died on March 29, 1921. At his funeral one admirer wondered what drew the large crowd of his friends to pay their last respects to this shy, retiring person. "I asked myself," wrote Dallas Lore Sharp, "what it was in this simple, childlike man, this lover of the bluebird, of the earth on his breast and the sky on his back, that drew these great men and little children about him. He was elemental. He kept his soul. And through the press men crowded up to touch him, and the virtue that went out from him restored to them their souls—their bluebird with the earth on its breast and the sky, the blue sky, on its back."

His friends formed the John Burroughs Memorial Association in his honor. This society set its major purpose of encouraging good writing in the field of natural science.

In his only book of poems, *Bird and Bough* (1906), Burroughs sang of the joys of nature in some three dozen selections, mainly centered around birds and the blooming of the forests during spring and summer. In "June's Coming," he wrote about an ecologist's dream of an unpolluted countryside at the dawn of summer. Here is one verse from that poem:

Again I see the clover bloom,
  And wade in grasses lush and sweet;
Again has vanished all my gloom
  With daisies smiling at my feet.

Burroughs' thoughts, penned two-thirds of a century ago, could serve us today as a guideline in our attempts to renew the harmonious ecological balance of times past.

10
Theodore Roosevelt

The first ripple of concern over conservation in America finally reached the White House at the turn of the century. After the frontier was closed in 1890, Teddy Roosevelt, on coming to the presidency in 1901, used his influence to preserve for future generations the wilderness wonders he had known in his youth. "Conservation" was the word he coined to call the attention of the citizenry to the new condition. His aim was to preserve—"conserve" it all—the clear streams, unspoiled skies and the wildlife resources, from the pigeon to the buffalo, that were vanishing from the land.

As the twenty-sixth president of the United States, Roosevelt established the Department of the Interior to ride herd on this task and, symbolically, its chosen emblem was the buffalo.

Of all President Theodore Roosevelt's varied activities both in and out of office—as soldier, statesman, writer, politician, and conservation naturalist—the last is the least well-known. Much has been written about his exploits in Cuba as head of his own volunteers during the Spanish-American War and his "trust-busting" but much less about his conservation programs, which many observers today consider the major achievement of his two terms in the White House.

Theodore Roosevelt was born in New York City in 1858 to a family of wealth and social position. As a boy he was thin, spindle-legged, asthmatic and nearsighted. To hold his own with other youths he took up track and boxing and was always a stubborn contender.

He developed an interest in natural history at an early age and became so proficient in taxidermy that his boyhood collection of birds and mammals numbered in the hundreds. He seriously considered a career in zoology, but poor vision and questionable lungs decreed otherwise.

At Harvard, which he attended from 1876 to 1880, young Roosevelt thought for a time that he might become a naturalist-adventurer like Charles Darwin. In his senior year he courted a young lady with such exuberant accounts of his "snakes and reptiles," which he kept in his college rooms in Cambridge, that he "frightened her out of her wits." But he gave up his naturalist bent for the more practical pursuits of law, although he never divorced himself completely from the lure of the wild.

After graduating from Harvard in 1880 he married his college sweetheart, Alice Hathaway Lee, settled in New York City and found his true calling in the rough-and-tumble of politics. A few years later he was struck a two-fold blow when both his young wife and mother died within twenty-four hours of one another. The shocked Roosevelt left the city and politics to manage his cattle ranch in the West.

In the quiet of the Badlands of Dakota Territory, he began his emotional reconstruction, driving cattle, hunting and writing, before returning East once again to take up his role as gentleman reformer amid a graft-ridden city and state. While riding the range with his cowhands, roping and rounding up cattle, he had not only restored his mental state but his physical health as well.

In 1886 Roosevelt returned to New York politics and resumed his climb to state and national prominence. Helping him in his colorful career was his second wife, Edith Kermit Roosevelt. As governor of New York Roosevelt began a reform administration that so disquieted the state's political leaders that they maneuvered him into the vice-presidency.

When he became president in 1901 on the assassination of McKinley, Roosevelt did not hesitate, in the face of strong political opposition, to use his power and influence to correct long standing abuses in the management of land, water, forests and wildlife.

Roosevelt's passionate interest in conservation of waterpower, protecting our national forests, reclaiming the arid Western lands by irrigation and protecting other dwindling natural resources could all be considered as part of his campaign against the so-called "robber barons." Before the turn of the century, oil and minerals found on public lands were bringing in enormous profits to a few at the expense of the general public. The same situation held true for much of our public grazing and timber lands as well as our waterpower resources.

Roosevelt's interest in conservation was always strong. As governor of New York, he emphasized the necessity of wildlife protection and secured legislation on forest preserves. As president he dwelt on the subject of national forests every year in his annual messages to Congress. His interest in this area along with the irrigation of desert lands and his opposition to the exploitation of waterpower was based on the heritage of his earlier years in the West combined with his conception (which was novel in that time) that these resources were the property of the people and should be used for their benefit.

In 1903 when the 57th Congress passed a bill awarding a certain N. J. Thompson the right to build a dam and construct a power station at Muscle Shoals, Alabama, the president vetoed the legislation with these strong words: " . . . the ultimate effect of granting privileges of this kind. . .should be considered in a comprehensive way and. . . a general policy appropriate to the new conditions caused by the advance in electrical science should be adopted under which these valuable rights will not be practically given away, but will be disposed of with full competition in such a way as will best substantiate the public interest."

After establishing the Interstate Commerce Commission, he went after the business trusts and prodded them to improve the health and safety of working conditions so that their employees and the general public could benefit.

His battles with the "selfish concentrations" of business tycoons were tied up intimately with his interest in conservation. As he laid his plans to conserve the natural resources of the country he was acutely aware that his main enemy was not the ravages of the weather and nature but the rapacity of the "despoilers of the earth for their private gain."

Roosevelt was aided in his administration by the advice and prodding of his chief forester in the Department of Agriculture, Gifford Pinchot. Together they established a conservation policy that reversed the former practice of allowing private interests to have their own way in the development of our natural resources.

In the area of waterpower conservation, Roosevelt put many of these vital areas under

Below: T. R.'s refusal to shoot a bear cub inspired this cartoon and the birth of the Teddy bear.

government control to prevent monopoly and speculation from running rampant. In March 1907 he created the Inland Waterways Commission to further this aim.

Roosevelt saw to it that cattle barons who were allowed to graze their herds on government lands were required to sign a lease and reimburse Washington for the privilege of using public land. Roosevelt furthered his multi-barreled approach to conservation by calling a White House conference of governors in May 1908 which resulted in the appointment of a National Conservation Commission headed by his friend, Pinchot.

This commission of 49 prominent leaders of science, industry and politics was named to prepare the first inventory ever compiled of the nation's prime natural resources and to recommend optimum uses of land. Seven months later a Conservation Congress met and suggested a plan to redeem lands that were either over- or under-watered.

Congress had already acted in this area when it dusted off Powell's plan for federal development of irrigation and passed the Reclamation Act of 1902. The first significant undertaking under this act was the building of Roosevelt Dam in Arizona to hold back the waters of the Salt River. The new lake and irrigation canals which were created by this engineering marvel helped turn a desert into one of the world's best farming areas. Twenty-five such irrigation or reclamation projects were started by Roosevelt.

Roosevelt's belief in conservation was based on a theory that the people were the real owners of public property and that the government, acting as their servant, should retain the ownership of these precious lands and waters for the benefit of all. His achievements in conservation are many and the policies and programs which he initiated still influence our thinking and action. During his two administrations more than 150 million acres of timberland were added to the national forests. The United States Forest Service was reorganized and a sound forest management policy was adopted. He brought Yosemite Park under national administration in 1906 and doubled the number of national parks when he added Crater Lake, Mesa Verde, Platt and Wind Cave parks.

When he first visited the Grand Canyon region in 1903 President Roosevelt voiced his impression of this phenomenon, which has served as a sort of Magna Carta to protect it from commercial exploiters ever since: "In the Grand Canyon," Theodore Roosevelt said, "Arizona has a natural wonder which, as far as I know, is in kind absolutely unparalleled throughout the rest of the world. I want to ask you to do one thing in connection with it in your own interest and in the interest of the country. To keep this great wonder of nature as it now is. . .leave it as it is. You can not improve on it. The ages have been at work on it, and man can only mar it. What you can do is to keep it for your children, your children's children, and for all who come after you as one of the great rights which every American if he can travel at all should see." In 1908, he designated Grand Canyon as a national monument.

In addition, fifteen other national monuments of unique scenic interest were dedicated during Roosevelt's eight years in office. More than fifty game preserves and bird sanctuaries in seventeen states and territories were created by executive order. This last series of moves was made without precedent, but no one challenged his authority.

From his boyhood hobby of bird-watching T. R. had grown and matured to become the most respected lover of nature—and particularly of birds—to occupy the office of the presidency. (The last letter he ever wrote, penned just a few hours before his death, was sent to William Beebe, the famed botanical zoologist and underseas explorer, asking him to explain a taxonomic point.)

He spent parts of each summer at his home on Long Island Sound with his wife Edith and his children. Surrounded by wide open fields and stretches of woodland, Sagamore Hill was a natural bird refuge. He catalogued which birds returned year after year to the same location, which ones were more abundant than the year before, and which were not so common as in his youth.

For example, when he returned to his favorite vacation retreat in the summer of 1906, he found two pairs of purple finches nesting there for the first time. His attention was drawn to them by "the bold, cheerful singing of the males, who were spurred to rivalry by one another's voices."

Later that same year, he called his sons' attention to a pair of rare red-breasted nuthatches making their peculiar "auk-auk" cry from a nearby elm. After viewing them, Roosevelt and his sons, Archie and Quentin, went inside their home on Sagamore Hill to look up the bird pictures in Wilson's ornithology. He considered this work and Audubon's as the most satisfactory ornithologies for the amateur bird student.

During the following summer of 1907, he noted that only one of the two pairs of purple finches returned to Oyster Bay and that the Baltimore orioles, for the first time in years, failed to hang their nests in the giant elm by the house. In their stead, he noted that black-throated green warblers took their place. He was so intrigued by this shift in nesting that he wrote the noted ornithologist, Frank Chapman, and asked him to visit Sagamore Hill for a day.

John Burroughs, the naturalist, did visit his home, and he and the president spent an entire day rambling through the meadow where Roosevelt had been cutting hay and the grove where he spotted the green warblers.

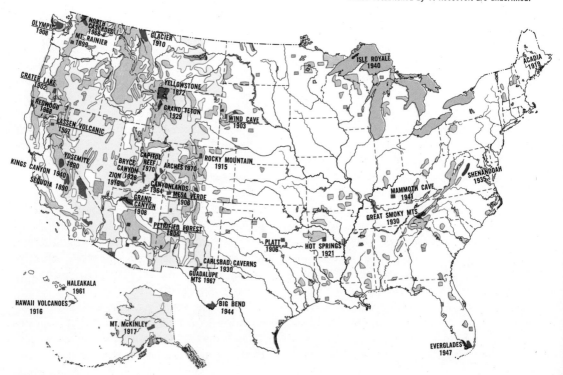

Conservation of natural resources first began in T. R.'s administration.

National Parks, Seashores and Monuments (Major)
National Forests and Grasslands
National Wildlife Refuges
Public Lands

National Parks are named with year of establishment. Those established by T. Roosevelt are underlined.

They observed several species of birds, from sparrows to finches and orioles, on their exploration of the grounds.

Roosevelt loved the out-of-doors and delighted in going on camping or boating trips with his family or friends. He often went on extended trips with his sons to the lonely eastern stretches of Long Island where they could become acquainted with nature in the raw. Even while president he continued to camp out with his sons, sleeping under the stars on a Navaho blanket, guarded only by his three boys, the eldest of whom was not yet fifteen.

The president did the cooking on these camping trips and his sons enjoyed his telling them stories after supper around a glowing fire. As our first naturalist-president since Jefferson, Teddy Roosevelt lived what he preached, by getting close to nature to relax from the tensions of the White House.

He seemed never too busy to record his observations and experiences bearing on the ecology and habits of animals, be they game species, songbirds or cold-blooded vertebrates. He wrote prolifically on conservation subjects for both popular magazines and technical journals. His books included such classics as *Winning of the West* and *Life Histories of African Game Animals*.

When his successor, President Taft, dismissed Gifford Pinchot, a passionate follower of Roosevelt, as his chief forester, it put T.R. in a bind since he was caught between two conflicting forces. Some observers have pointed out that Pinchot did Roosevelt and the cause of conservation a distinct disservice by forcing Taft to dismiss him, since this act also accelerated the quarrel between Taft and Roosevelt.

Top: Roosevelt, the outdoorsman, in the Rocky Mountains.

Center: The President cooking over an outdoor fire on an outing on Long Island.

Bottom: Roosevelt on his expedition to the Amazon jungle.

Rebuffed by the electorate in 1912 after running for the presidency for a third time on a Progressive-Bull Moose ticket, the restless Roosevelt took off for the jungles of Brazil. Back to nature once again, he enjoyed the thrill of exploring an unknown tributary of the Amazon. While on the expedition an injury to his thigh and an attack of jungle fever laid him out. For a time it appeared that he might not get back to civilization. Arriving back home, his body weakened but his eyes still flashing, the old Rough Rider spent his last years at his beloved Sagamore Hill. On January 7, 1919, Theodore Roosevelt died peacefully in his sleep at the age of sixty.

His spacious home on Oyster Bay, Long Island, New York, is now a National Historic Site. It is within a mile and a half of the Roosevelt Bird Sanctuary, administered by the National Audubon Society as a symbol of his lifelong interest in the conservation of wildlife.

Out in the Black Hills of South Dakota, the late sculptor, Gutzon Borglum carved four heroic profiles of our leading presidents into the granite hillside. Nestling shoulder-to-shoulder with Washington, Jefferson and Lincoln is Theodore Roosevelt, who, while he might not go down in history as one of our greatest chiefs of state, certainly must rank as the most concerned conservationist president. For that reason alone he richly deserves a high place among our nation's heroes. In the opinion of the Progressive Senator Robert La Follette of Wisconsin, who had become one of his staunchest allies in the conservation push, this facet of his presidency would stand as Roosevelt's greatest work. La Follette said that the chief executive had started a "world movement for staying territorial waste and saving for the human race the things. . .on which alone a peaceful, progressive and happy life can be founded."

11
Gifford Pinchot

Before 1907 the word "conservation" was not yet known or used in the American lexicon. The modern conservation movement unquestionably owes much of its impetus to former Pennsylvania Governor Gifford Pinchot, who credits two friends, Overton Price and W. J. McGee, with first suggesting the term and the need to take action. Pinchot gave to the new movement his personal zeal and dynamic flair for publicity, which brought into our national thinking the idea of "making available to the people as a whole the God-given resources of the earth."

His dedication to conservation has been often underestimated, but his capacity for inspiring dedication among his subordinates helped him in his strenuous pursuit of conservation policies. Before 1900 any advocate of conservation of natural resources was like a prophet crying in the wilderness. Theodore Roosevelt made conservation a national issue in 1901 and appointed Pinchot as head of the nation's forest reserves. For the first time restrictions were placed on federal lands, preventing the unlimited exploitation of their timber, minerals, waterpower sites and wildlife.

Together, Teddy Roosevelt and Pinchot served as the best hard-hitting team that American conservation ever had. Although a distinguished group of mountain men, poets, naturalists and statesmen had anticipated their concern for the preservation and systematic use of America's natural resources, Theodore Roosevelt and Pinchot possessed both the vision and political power to act. History shows that these two qualities of leadership make a winning combination, and in this case it worked for the betterment of the nation.

Theodore Roosevelt and Gifford Pinchot stopped the unchecked activities of "robber barons" whose greed and destruction ran counter to the democratic idea. Roosevelt had observed that such democratic license had triggered a "riot of individualistic materialism, under which complete freedom for the individual. . .turned out in practice to mean perfect freedom for the strong and none for the weak."

In the beginning of this nation's history there were some 850 million acres of virgin forest. By 1920 all but one-fifth of that land had felt the sharp cuts of the woodmen's axes. The rapacious consumption of this vital natural resource by the timber barons was staggering. These stark figures reflecting the great forest raids of the latter part of the nineteenth century show the state of affairs which Pinchot faced when he took office under Roosevelt.

Gifford Pinchot was the right man in the right place at the right time to stop this onrush of man's folly in eliminating the earth's oldest living things for selfish ends.

Pinchot was born to a prosperous family in Simburg, Connecticut, on August 11, 1865, at the close of the Civil War. After attending Phillips-Exeter Academy and graduating from Yale in 1889, he went abroad to study forestry (then known as silviculture) in France and Switzerland. Conservation had been practiced to some extent in Europe for centuries. Diminished wood supplies had prompted decrees restricting cutting on certain lands as early as the 13th century. Upon his return to the United States, Pinchot made the first systematic study of forestry in this country while managing George Vanderbilt's Biltmore estate in the Great Smoky Mountains of North Carolina during 1892.

The general purpose of Pinchot's early work was to develop methods of efficient and profitable timber production. Forest preservation, he learned, was not incompatible with the profitable use of forests. He soon proved this by encouraging selective cutting and by thinning forests of dead and diseased wood, scrub growth and mature trees. He established the principle of "constant annual yield" and instructed the forest crews at the Vanderbilt estate in methods of cutting trees without destroying the forests.

With that background, it was only natural that Pinchot was selected as chief conservation advisor when Theodore Roosevelt assumed the presidency in 1901. Pinchot soon became his guiding light and eventually his close confidant in a wide range of high-policy matters.

Stewart Udall called Pinchot a "magnificent bureaucrat" because his talents as a scientist, politician and publicist blossomed in Washington. He was able to sell his and Roosevelt's conservation and forest management policies on Capitol Hill as no one else had been able to do.

He arranged conferences and legislation to achieve Theodore Roosevelt's goal and, before he was through, 132 million acres of forest land had been declared a public domain to be set aside and scientifically managed as national forests.

During the period of Roosevelt's administration, the entire Forest Service support and administrative machinery was set up. Pinchot served as chief of the Forest Service from 1898 to 1910. His enthusiasm and personal publicity did much to aid the conservation movement to save our precious timberlands.

Besides his main duty, he also found time to serve on the Committee for Public Lands (1903) and the Inland Waterways Commission (1907). He held the chairmanship of the National Conservation Commission (1908) and the presidency of the National Conservation Association (1910).

Somehow he managed during this period to help found the Yale School of Forestry and in 1903 became a professor of forestry there. He wrote two books on the subject: *Primer of Forestry* (1899) and his classic, *The Training of a Forester* (1917), which went through four editions.

Destructive logging on public land in Colorado. Land has been cut clean for mine props.

In 1910 Pinchot lost his position as the nation's chief forester when he was fired by Theodore Roosevelt's successor, President Taft, in a nationally debated controversy over federal conservation policies. He had fallen out with Taft in a dispute over the permissive land use policies of the president's Secretary of the Interior, Richard Ballinger. He publicly charged Ballinger with allowing private acquisition of vast coal lands in Alaska. Pinchot also clashed with conservationists who did not share his strong belief that scientific forestry should be practiced on *all* forest lands.

He considered scenic preservation and the natural wilderness to be of secondary importance. "Forever wild" meant to Pinchot forever mismanaged. When the State of New York prohibited cutting from its state forest preserve in order to keep a portion of its forest land in a natural state for the spiritual and recreational enjoyment of future generations, Pinchot lamented at the time that New York had "vetoed forestry."

In 1904 Pinchot became embroiled in a bitter controversy over a portion of Yosemite Valley—the Hetch-Hetchy basin. His chief opponent and long-time friend, John Muir, the naturalist and founder of the Sierra Club, took issue with Pinchot's conservation-for-use theories. A battle royal ensued and, although Muir lost that skirmish, he did not lose the war.

In retrospect, Pinchot's stubborn insistence that unused land is wasteland has diminished his historical importance in the conservation movement. Although a more balanced government land policy is now in vogue today, Pinchot's present-day critics would be the first to acknowledge that if it weren't for this "magnificent bureaucrat," the issue would no longer be either relevant or debatable.

After 1910 Pinchot constantly urged Roosevelt to challenge Taft and the Republicans, which made him a "pest" and a "radical" in the eyes of the hero of San Juan Hill. Although he often received a cold shoulder from Theodore Roosevelt, Pinchot still looked upon the conservation of our national resources and their protection against exploitation by selfish interests as an overwhelmingly important national issue. He worked constantly to push Roosevelt into a battle with Taft over this item of national welfare. For the rest of his life Pinchot served as a public gadfly and crusader-at-large for conservation.

When Theodore Roosevelt bolted the Republican party and ran as a "Bull Mooser" in his unsuccessful campaign of 1912, Pinchot helped Roosevelt to carry Pennsylvania, where he won over Taft and Wilson by a wide margin. In 1914 Pinchot was the Progressive party's candidate for a seat in the U.S. Senate but he was defeated.

Pinchot became more active in Pennsylvania's conservation affairs. A solid foundation had been achieved before 1900 with the establishment of a state game commission, a fish commission and a forestry commission. Pinchot sought to restore in the Keystone State the game, fish, streams and forests to a condition approaching that of the time of William Penn. The often irascible Pinchot

> Pinchot called for wise use of our timber resources through the creation of "national forests."

was just the man to lead the fight, despite the charges hurled at him by his Republican organization opponents that he was a "Socialist" and a "Communist." In 1920 he was appointed commissioner of state forests.

In his first successful gubernatorial campaign, running as a Republican in 1922, Pinchot did not emphasize conservation but rather the need for a "new order" in state government. After assuming office, however, he moved fast to bring together scattered functions of the Commonwealth government into a new Department of Forests and Waters. Progress was made in forest conservation and the first steps were taken to use the forests for public recreation. A Power Survey Board was also created by Pinchot to study the water and other power resources available for industry, transportation, utilities and other needs.

His board's recommendation provoked powerful dissent from the chief financial interests in the state, since the recommendation proposed a scheme for the strong regulation of public utilities. The implementing legislation met with defeat and Pinchot never achieved his desired goal in this area. Nevertheless he did secure a strengthening of laws relative to stream pollution and public health.

Pinchot was progressive and democratic in his philosophy but autocratic in his personal life. In trying to imitate his hero, Teddy Roosevelt, Pinchot chose to exhibit his personal vigor by reviewing the National Guard on horseback or, as governor, riding throughout the state in an open touring car—no matter how inclement the weather.

As a man who confounded his friends as well as his enemies, the impressive-looking, white-mustachioed Pinchot was either loved or hated. His flowing mustache, flying in the wind, gave rise to his nickname "Old Handlebars." Gifford Pinchot was one of the most colorful, progressive and capable figures ever

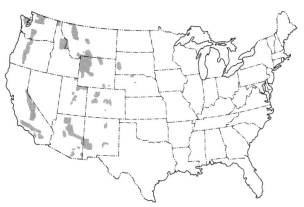

NATIONAL FORESTS 1900

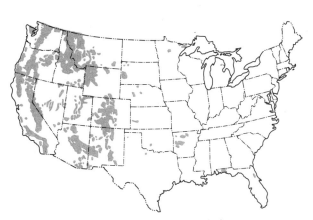

NATIONAL FORESTS 1910

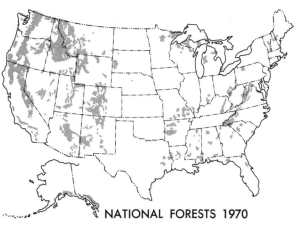

NATIONAL FORESTS 1970

to be governor of Pennsylvania.

Since as governor he could not succeed himself, Gifford Pinchot was denied an immediate opportunity for a second term. He left office with a farewell address in January 1927 in which he expressed the "most hearty contempt not only for the morals and intentions, but also the minds of the opposing politicians of Pennsylvania," including the powerful Mellon family of Pittsburgh and the Vares of Philadelphia. He even helped to deny a U. S. Senate seat to William Vare when he sent a note to the Senate stating that Vare's "nomination was partly bought and partly stolen."

On the advice of lieutenants in Harrisburg that the conservation and control of utilities issues were still alive Pinchot sought the governorship again in 1930. A well-publicised trip to the South Seas, a book of his and the subsequent movie called *To the South Seas*, describing his adventures on the ship *Mary Pinchot*, contributed to his reelection.

Despite the loss of the powerful Mellon backing in the Republican primary, Pinchot stumped the state with his wife, using his usual hammer-and-tongs style. He argued for an improvement and an enlargement of farm life through better rural roads as well as for stronger regulation of the utilities. He won the primary and beat his Democratic opponent in the fall elections in 1930 with a plurality of 59,000 votes.

Upon assuming the governorship for the second time, Pinchot moved in on the powerful utilities and, with the support of a friendly Assembly, he eventually won control of the State Public Service Commission which regulated these institutions.

In the early 1930s the specter of the Great Depression stalked Pennsylvania and the nation. By the time Pinchot took office one-fourth of the state's labor force was out-of-work. When statewide unemployment reached a peak of 1,132,000 in 1932, Pinchot tried to act to "conserve" the welfare of these citizens, but the state legislature refused to accede to his demands which were put forward in three special sessions of that body. The majority was not yet willing to admit that the time had come for large-scale state relief aid. Finally, a State Emergency Relief Board was authorized to supplant direct appropriations to local welfare boards.

The most significant accomplishment of his second term was the building and paving of 20,000 miles of rural dirt roads. Although these macadam roads, which became known as "Pinchot Roads," were narrow and lacked modern engineering know-how, they did take the farmer out of the mud. In 1933 he also urged old-age pensions, compulsory health insurance and minimum wages for women and children long before they were adopted in the rest of the nation.

In his retirement Pinchot penned a lively autobiography, *Breaking New Ground*, that was published a year after his death in 1946. In that book Pinchot correctly posed the conservation dilemma facing the United States at the turn of the century. "The American Colossus," he observed, "was fiercely intent on appropriating and exploiting the riches of the richest of all continents, grasping with both hands, reaping where he had not sown, wasting what he thought would last forever. New railroads were opening new territory. The exploiters were pushing farther and farther into the wilderness. The man who could get his hands on the biggest slice of natural resources was the best citizen. Wealth and virtue were supposed to trot in double harness."

Because he dedicated himself to ending such exploitive attitudes toward natural resources, Pinchot was a key figure in establishing truly conservationist policies on a federal governmental level.

12
George Washington Carver

After the rich Southern plantation owners had ruined their cotton fields by overplanting year after year for over a century, it took a small Black man, born a slave in the Civil War, to bail them out. His radical ideas—the reconstitution of the ravaged land through crop rotation and the planting of the lowly peanut, soybean, cowpea, and sweet potato—brought the land back to life.

Carver's innovations and discoveries eventually freed the South from the tyranny of King Cotton by restoring the essence of life to millions of spent and sterile acres. Carver was our first soil conservationist. It was ironical that those Whites who refused to call him "Mister" profited from his genius even more than the impoverished Blacks of the Old Confederacy. Subjected to racial discrimination throughout his long life, Carver rose above it and never showed resentment, continuing to serve all mankind.

Carver was born on a Missouri farm (probably on July 12, 1861) to a slave girl named Mary, who had been purchased by a hardworking White farmer named Moses Carver. As was the custom, slaves took their master's surname, so the young boy became known as Carver's George. Born sickly and given little chance to survive, he had to struggle to overcome his handicaps.

Denied the opportunity to attend an all-White elementary school in nearby Locust Grove, young Carver, then aged fourteen, persuaded his master to let him go to Neosho, the former Confederate capital of Missouri, to attend a segregated colored school. The year was 1875, and the frail, stuttering Carver stumbled into the yard of a hardworking, childless Black washerwoman named "Aunt" Mariah Watkins, who took him in and gave him shelter and food. She had him change his name from Carver's George to George Carver and encouraged him in his quest for education.

He became the most apt pupil in the one-room, one-teacher schoolhouse that was packed daily with seventy-five Black children. He had already developed a consuming interest in studying the peculiarities of seeds and flowers. In reaction to a lynching at nearby Fort Scott where he next attended school Carver wandered around the midwest for the next ten years, attending dozens of schools—each for a short period of time. He earned his keep by cooking or doing laundry.

Now in his early twenties, he joined a group of migrants in the foothills of the Rockies and picked fruit with them all the way to New Mexico. One morning in the desert he sketched a beautiful plant with needled spears and a pale waxen plume on a crumpled scrap of paper. Fifteen years later, his painting of the *Yucca Gloriosa* won a prize for Carver at the World's Columbian Exposition in Chicago.

Settling down in Minneapolis, Kansas, he was once again befriended by another Black washerwoman, Lucy Seymour, who was also to have a profound influence on his life. After graduating from high school he enrolled in small, Presbyterian-run Highland College in northeast Kansas in the fall of 1885 to find answers to the mysteries "of what makes it rain and what makes the sunflowers grow."

On arriving at the college he was turned down by the principal, a Reverend Brown, because "You didn't tell me you were a Negro. Highland College does not take Negroes." This blow did not deter the saddened and frustrated George, who was now past twenty-five. In 1886 he homesteaded a 160-acre tract in Kansas and built himself a sod house on his land. He even kept flowers alive during the harsh Kansas winters in a lean-to conservatory which he built on the south side of his modest home.

Two years later he mortgaged his home and walked back East, stopping at Des

Moines where he was befriended by a White medical doctor who helped him to enroll in Simpson College in Indianola. This school of about 300 students fortunately accepted Blacks. Carver was the second Black student to enroll there.

At first he thought he wanted to be an artist, since he loved to paint flowers and plants. He was encouraged by his art teacher, however, to transfer to the Iowa State Agricultural College at Ames. Because of his race she felt he could make a better living from agriculture than art. Since agricultural research was becoming an important science at the turn of the century, Carver made the move at the right time when he left Simpson and began study at the agriculture school at Ames.

In the laboratories there Carver became acquainted with the magic of chemistry. He learned about the nutritive requirements of the soil and plants and new methods of soil and crop testing, although he at first had to suffer the humiliation of eating his meals in the college cafeteria basement with the field hands at the agricultural experiment station.

He obtained his B.S. in 1894 near the top of his class with a thesis entitled: *Plants as Modified by Man*. He was the first Black man ever to be graduated from Iowa State.

Well past thirty upon graduation, he was offered the position of assistant botanist in charge of the college greenhouse, and it was then that he acquired his first taste of the importance of conservation. "Nations last only as long as their topsoil lasts," said one of his professors and Carver now began an intensive study of the interrelationships between plants and earth. At this time he also started to study for his M.A., specializing in mycology—the study of fungus growths. His collection soon numbered over 20,000 specimens. Through hybridization he produced large numbers of fungus-resistant plants and fruit.

Soon he found himself lecturing all over the state to farmers and county agricultural agents. "If you took an Iowan to the North Pole and left him there, it would be necessary to provide him with food and warm clothing or he would perish. In the same way, you cannot plant an apple tree in alien ground and expect it to flourish without special care," he said.

In 1896 he obtained his master's degree in bacterial botany. Carver felt that it was now God's plan that he should turn his knowledge back to his people.

At this time Booker T. Washington was struggling to keep alive his fifteen-year-old Tuskegee Normal and Industrial Institute. Although he was the most noted Negro in America at the time, he had to fight off financial disaster for his school for Blacks. He had no one on his staff with any knowledge of agricultural science and, when he heard of Carver he wrote him a historic letter on April 1, 1896: "I cannot offer you money, position, or fame. The first two you have. The last, from the place you now occupy, you will no doubt achieve. These things I now ask you to give up. I offer you in their place work—hard, hard work—the task of bringing a people from degradation, poverty and waste to full manhood."

So in 1896, a few months after receiving the letter, the tall, hawk-faced Carver journeyed from Iowa State to south-central Alabama to join Washington at the Tuskegee Institute. There he stayed until his death forty-seven years later, becoming more than a scientist and doctor of plants. He forsook all offers of salary increases and overcame bitter obstacles to become a benefactor of his race and all mankind. Significantly, at his death in 1943, he was still earning the same $125.00 per month salary that he started with in 1896.

The first obstacle was the lack of equipment and facilities at the institute. Washington told him, "Your department exists only on paper, Carver, and your laboratory will have to be in your head." His annual salary at Tuskegee was lower than his Iowa salary, but he accepted it because he deeply believed that "this line of education is the key to unlock the golden door of freedom to our people..."

It did, and in the next half century Carver helped to spread the message of Tuskegee around the world. There were no two more diverse men than Carver and Washington, though they worked for the same common goal—the uplifting of their own oppressed people.

The head of Tuskegee knew that the future growth of his dream depended on keeping Carver happy and removing as many frustrations as possible from his path of experimentation in finding new ways to conserve the soil. Washington backed up Carver when the latter complained about the interference in his work by Booker's older brother John and the "office people" at the institute.

In the spring of 1897 Carver took twenty acres of ravaged, gullied sandy soil at Tuskegee and began a small experimental farm. He scavenged equipment from the town dump and back alleys to help him test soil samples in his laboratory.

"They told me it was the worst soil in Alabama," Carver commented, "and I believe them. But it was the only soil I had. I could either sit down and cry over it or I could improve it." He chose the latter, and with a donation of fertilizer from an Atlanta company and organic waste Carver transformed the starved land into an oasis in a few years' time.

He had thirteen student farmers who eventually went out and became missionaries to poor illiterate sharecroppers and tenant farmers. One of Carver's pupils, Tom Campbell, became the first Negro field agent in the

Professor Carver's demonstration wagon made his laboratory work available to poor farmers.

U. S. Department of Agriculture.

In the fall of the first year that Carver arrived, the twenty-acre plot yielded 5 scrawny bales of cotton and 120 bushels of sweet potatoes for a net annual loss of $16.50.

By the second year, after planting cowpeas to give nitrogen back to the soil, the land began to show a profit. When he finally replanted cotton in the now-enriched soil, it yielded unheard of amounts that astounded both Black and White farmers.

Although Booker T. Washington was later criticized as an "Uncle Tom" for bowing to the White man and only training the Black students who came to Tuskegee for menial, semi-skilled jobs, this charge rarely was applied to Dr. Carver. His multiple contributions helped uplift the agricultural sciences for both Black and White and indirectly provided the spark that set off a wave of national soil conservation and crop rotation practices that were to be his legacy.

When Booker T. Washington died in 1915, former President Theodore Roosevelt came to Tuskegee for the funeral and told Carver: "There is no more important work than what you are doing." Coming from our first conservationist president, this was a high tribute indeed, and Carver cherished it.

Carver, with his simple and rational approach to problems, was not just interested in conserving the land but was equally concerned in saving the people. Wherever he found an opportunity to help poor farmers rise up from their poverty-stricken lives, he made humble suggestions that sooner or later would find acceptance in Macon County and other areas surrounding the institute.

Saving hog fat and converting it into soap, filtering yellow wash paint from the Alabama clay hillsides to spruce up the weather-beaten shacks and removing the curse of pellagra by supplementing diets with vegetables were three of Carver's contributions.

His movable school, in the form of a rickety mule-drawn wagon driven by Carver and his assistant, Tom Campbell, instituted a new method of bringing help to the farmers where they were. This idea spread to dozens of foreign countries and Carver believed it was his most important contribution.

Carver once said that he loved "to think of nature as an unlimited broadcasting system through which God speaks to us every hour, if we will only tune Him in." This attitude toward nature motivated him as a central driving force during his long and dedicated life.

Carver believed that man's civilized progress fell into three phases: first, finding raw materials; secondly, the adoption of these natural fruits to make food, metals and fibers; and thirdly, the creation of new substances by making chemical changes in substances already in existence.

"I believe the Creator has put ores and oils on this Earth to give us a breathing spell," he said. "As we exhaust them, we must be prepared to fall back on our farms, which is God's true storehouse and can never be exhausted. For we can learn to synthesize materials for every human need from the things that grow."

Many believe that this sweeping concept of Carver's was his most important and enduring contribution to mankind.

With his keen vision, Carver had been urging since 1899 a reforestation program to prevent erosion and to assure adequate timber for future generations. But it wasn't until 1932, after a series of disastrous floods, that such a national program was officially considered. When the Great Depression left the agricultural South in a state of prostration, it was Carver's ideas that once again helped guide the federal government in institution policies to save the farmers and the land with soil bank conservation methods.

Below: Carver bulletins (44 were published) contained solid information on such things as soil building, using wild plums and acorns.

No one in his day knew more about the chemical magic of plants and how best to use this knowledge for man's good than George Washington Carver. Working in his laboratory, Carver extracted particles of matter from plants and combined these particles into new foods, building materials and medicines. He discovered over 300 uses of the peanut—running from plastics, food, oils and milk, and 107 uses of the sweet potato. These two plants, which provided a practical alternative to cotton and its scourge, the boll weevil, became two of our foremost agricultural products. He made from the peanut such unlikely things as milk, dyes, cosmetics, flour, paper, synthetic marble and soil conditioners. An accomplished painter who ground his own colors from native Alabama red clays, he then donated them to friends. He even turned down an offer from Thomas Edison to work with the great American inventor at an unheard of salary of $100,000 a year.

In 1940 when President Franklin D. Roosevelt visited the Tuskegee campus, he asked to see Carver. "You are a great American, Professor," he told the stoop-shouldered, gnarled old man. "What you have done in your laboratory has made the whole nation stronger."

Also a visitor to Tuskegee was Vice-President and once Secretary of Agriculture Henry Wallace, a leading hybrid seed fancier. Wallace, who remained his friend for over half a century, was first introduced into the mysteries of seed fertilization by Carver while at Iowa State Agricultural College. "This scientist, who belonged to another race, had deepened my appreciation of plants in a way that I could never forget," Wallace said later.

George Carver never spoke of death despite his desperate struggle with anemia for several years. Sometime in the early evening of January 5, 1943, he said to his nurse, "I think I'll sleep now," and quietly passed away a few hours later. He died with no known relatives, but hundreds of mourners came to view his body as it lay in the chapel.

Although we were caught in the middle of World War II, Senator Harry Truman of Missouri introduced a bill providing for the establishment of the George Washington Carver National Monument on the site of the original Moses Carver farm near Diamond Grove, nestling in the Ozark foothills. After passage by Congress, the farm and the woods and fields that Carver walked as a boy became part of the National Parks system.

The epitaph on his tombstone is befitting of his contributions to mankind. It reads: "He could have added fortune to fame, but caring for neither, he found happiness and honor in being helpful to the world."

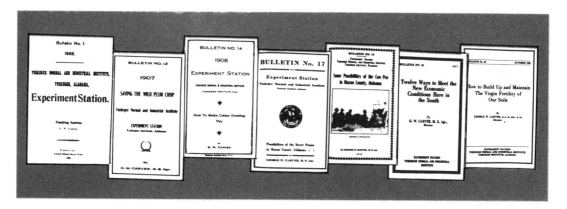

13
Rachel Carson

It was Eric Sevareid, the eminent TV commentator who compared the late artist, poet, scientist, author and reformer, Rachel Carson, with Harriet Beecher Stowe. He felt that as the latter nineteenth-century writer had helped touch off the Civil War with her classic, *Uncle Tom's Cabin*, so her twentieth-century counterpart has sparked a war against pesticides with her equally controversial book, *Silent Spring*.

Her disarmingly eloquent book, published in late 1962, was a two-fisted indictment of our devotion to modern technological innovations which push nature to yield more and more and more without a compensating loss. The *New York Times* felt at the time that if *Silent Spring* "helps arouse enough public concern to immunize Government agencies against the blandishments of the hucksters and endorse adequate controls, the author will be deserving of the Nobel Prize as was the inventor of DDT."

While she lived, Miss Carson received neither the Nobel Prize nor approval from the government for her efforts, but she left a noble legacy, which helped to usher in the age of ecology a few years after her untimely death.

Rachel Carson was a rare blend—a physical scientist who was also a gifted writer. Her poetic writing about nature caught the imagination of the general reader. She showed this talent with her three books on the ocean. In 1951 her classic work, *The Sea Around Us*, received the coveted National Book Award.

Rachel Carson was born in Springdale, Pennsylvania, a small town lying northeast of Pittsburgh, on May 27, 1907. Her mother taught her to appreciate the natural life abounding in this small suburban community, which led the young girl to her future career and worldwide fame. She fell in love with the birds and insects that came and went in the surrounding countryside. After graduating from a local high school, she enrolled in Pennsylvania State College for Women in Pittsburgh, with the intention of pursuing a writing career.

While taking a course in biology there, she renewed her latent scientific inclinations, in time deciding to take her degree in biology. After graduating in 1929 she went on to Johns Hopkins University in Baltimore for postgraduate study. In 1931 she started her teaching career there, and a year later became a member of the zoology staff at the University of Maryland, where she remained for the next five years.

She never married, claiming that she didn't have time, but she did find time between writing and work to devote to her aged mother and an adopted nephew.

This introverted, gentle biologist spent fifteen years working quietly in the laboratories of the federal government, writing and editing publications for the U. S. Fish and Wildlife Service. In the 1950s she achieved literary fame for her classic works on the oceans: *Under The Sea Wind, The Sea Around Us,* and *The Edge of the Sea*. But it was her fourth and last book *Silent Spring*, which marked her as a leading conservationist and crusader for the preservation of the natural environment from man's encroachments.

In this pivotal work she narrated the great harm done to insects, birds, animals and fish by the pesticides and insecticides which were used indiscriminately by farmers and planters to increase the yield of their crops. The resulting controversy which this volume engendered shook up the departments of Agriculture and Interior as well as the chemical industry which produced dangerous chlorinated hydrocarbon products including DDT.

When, in the mid-1950s, she first became aware of the dangers of pesticides, DDT in particular, through correspondence with friends, Miss Carson tried to get some other well-known author-friends to write on the subject. She didn't feel up to it herself and tried to convince E. B. White of *The New Yorker,* among others, that he was better fitted to accomplish the task.

In early 1958 White wrote in reply that he couldn't take on the assignment, but stated "I think the whole vast subject of pollution... is of the utmost interest and concern to everybody. It starts in the kitchen and extends to Jupiter and Mars. Always some special group or interest is represented, never the earth itself." This letter established the course of Rachel Carson's remaining years and her legacy to history.

During her four-year struggle to finish *Silent Spring* she rewrote her manuscript many times in order to pare it down to a short, simple exposition that would appeal to the American public rather than to a few academic scientists. The completion date was postponed several times between 1959 and mid-1962. While working on her task Rachel Carson discovered that she had cancer. This knowledge gave her an added impetus to finish the book before it was too late.

She wrote hundreds of letters to governmental, industrial and university experts as well as to friends in the pursuit of concrete information on pesticides and its effects, keeping their replies—her "ammunition," as she called it—to herself until she was ready to unleash her time bomb on the nation.

In *Silent Spring* she warned us of the contamination of earth, water and air by an unchecked "rain of pesticides." She amassed a crushing amount of evidence, most of it solidly scientific, to hammer across this point: "Along with the possibility of the extinction of mankind by nuclear war, the central problem of our age has become the contamination of man's total environment with such substances of incredible potential for harm—substances that accumulate in the tissue of

Dust cover for the book which aroused the public to the dangers of needless chemical pollution.

plants and animals and even penetrate the germ cells to shatter or alter the very material of heredity upon which the shape of the future depends."

The future world, Miss Carson warned, is in danger of falling silent when birds, fish and other wildlife succumb to the poisons man keeps pumping recklessly into his environment. She sought to explain the delicate balances of the natural world, and she implored her fellow man to take notice of his folly.

The chlorinated hydrocarbon compounds used in killing sprays for insect pests are extremely durable and move up the food chain in a fashion never anticipated. In a typical example of DDT spraying over a mosquito-infected lake the poison was quickly absorbed by tiny water plants (zooplankton) who in turn were fed upon by fish with a subsequent increase in the concentration of DDT. Birds such as the eagle or falcon which fed on the fish were affected in their egg-laying process. The birds' eggs lacked

sufficient calcium and protective strength. The end result—few chicks hatched and in time, the population of a local nesting ground was erased.

Silent Spring took four years to research and write, but it took twice that time period for the government bureaucracy to react favorably to her recommendations. Her charges that pesticides could have detrimental side effects on human life united powerful industrial and governmental interests against her in what might be called the "agricultural-chemical complex."

From the business world, the list of corporations who financed the attack on *Silent Spring* read like the chemical industry's representation on *Fortune* magazine's top 500 companies. On the government side, the Department of Agriculture had previously established a hard-line goal of exterminating with insecticides all species of insects that might damage industrial crops. One USDA officer claimed that current safeguards were adequate and that no additional laws were necessary to protect the public from abuses of these products. The agricultural-chemical complex attempted to discredit Rachel Carson by suggesting that the *only* alternative to massive pesticide use would be crops plagued by hordes of insects, with the resulting specter of worldwide famine.

Even before the book was published the chemical industry unleashed an intense public relations attack against her charges. In July 1962 the *New York Times* ran a meaningful headline: "SILENT SPRING IS NOW NOISY SUMMER," the newspaper described the controversy developing among governmental, agricultural and chemical experts over the serialized and shortened Carson work which appeared in *The New Yorker*. This skirmish foreshadowed the storm which broke over her head and her book when it finally appeared between hard covers in September.

One California chemical corporation president said sarcastically that Rachel Carson wrote not as a "scientist but rather as a fanatic defender of the cult of the balance of nature." Another critic called it a "hoax" and others hurled abuse at her without even reading the book, wherein Miss Carson pointed out that she was not calling for the abandonment of *all* pesticides, but just those long-lasting chlorinated hydrocarbon products like DDT.

Time denounced her book as "an emotional and inaccurate outburst" and accused her of "putting her literary skill second to the task of frightening and arousing readers." Another personal attack linked her with "food faddists and health quacks."

Miss Carson hit back at her critics with a rebuttal in the *Audubon Magazine* citing many examples of the harm done to nature by pesticides. She was amused to read a bit of wishful thinking in one of the trade magazines. "Industry can take heart," it said, "from the fact that the main impact of the book, *Silent Spring*, will occur in the late Fall and Winter—seasons when consumers are not normally active buyers of insecticides . . . it is fairly safe to hope that by March or April *Silent Spring* no longer will be an interesting conversational subject." How wrong they were in their estimation of this book's long-range impact.

She rightfully pointed out that the vitriolic campaign to discredit her was not "merely a campaign against an author and a book" but was "a campaign against a cause, and that cause was the promotion of sanity and restraint in the application of dangerous materials to our environment."

Then, pinpointing her charges against the selfish chemical industries manufacturing insecticides, she asked the simple question: "As you listen to the present controversy about pesticides, I recommend that you ask yourself: Who speaks? And why?"

Below: Diagram of how a toxic substance becomes concentrated as it moves through the food chain.

Far right: Audubon Medal awarded to Rachel Carson in 1963.

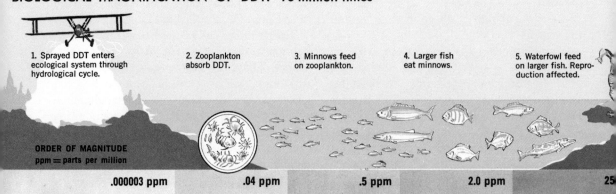

Silent Spring finally led to the formation of a high-level White House sponsored President's Science Advisory Committee on Pesticides. This group wrote a report commending Miss Carson's services to the nation. It also led to continuing studies on the abuses of insecticides by special committees of the House and Senate.

In November 1969 the U. S. government vindicated Rachel Carson's campaign against the unwarranted use of pesticides by banning the most infamous one, DDT, partially at first in 1970-1971 and absolutely by 1972. The U. S. government found that DDT not only polluted the waters of the country and killed the fish and wildlife but was a definite threat to the health of human beings who ate food sprayed with the product. The lone fighter against the monster of dichloro-diphenyl-trichloro-ethane (DDT) had won a battle in death that she had started in life.

If she accomplished no other end, she helped quietly to educate the populace on the potential hazards of the indiscriminate use of pesticides.

She took a Thoreau-like stand when she told a congressional investigating committee in 1962 following the publication of her book, that "I deeply believe that we in this generation must come to terms with nature." Her simple prophecy and challenge could easily become a clarion cry to us, if we want to insure that the coming generations survive on this fragile planet.

Although many persons, including several eminent scientists, tried to discredit her as being an overly hysterical alarmist, less than a decade later Miss Carson is largely credited with having been the voice of reason that ushered in an era of environmental awareness.

On December 3, 1963, Rachel Carson was awarded the coveted Audubon Medal for her gallant effort to arouse the public through her book *Silent Spring*. This event marked a high point of her career since it signified that her colleagues in conservation had singled her out for acclaim. "It took the force and clarity of Miss Carson's gifted pen to bring about the first genuine reappraisal by government and by the public of the strange new poisons that have been poured into the environment all about us," said Carl Buchheister, the National Audubon Society's president.

In her sensitive acceptance speech, the first woman ever to be honored with the society's highest award impressed on her audience the urgency of the crisis that America

faced. She made the following points:

"Over the decades and the centuries, the scenes and the actors change. Yet the central theme remains—the greed and the shortsightedness of the few who would deprive the many of their rightful heritage.

"If the crisis that now confronts us is even more urgent than those of the early years of the century—and I believe it is—this is because of wholly new factors peculiar to our own time.

"These are, first of all, the phenomenal growth of the human population, threatening to over-run its own environment in a way that can bring only deep concern to thoughtful students of population problems.

"The second factor is a corollary of the first: that as people and their demands increase, there is a smaller share of the earth's resources for each of us to use and enjoy. There is less clean water, less uncontaminated air; there are fewer forests, fewer unspoiled wilderness areas.

"The third reason is the introduction of new and dangerous contaminants into soil, water, air, and the bodies of plants and animals as our new technology spreads its poisons and its discarded wastes over the land."

Rachel Carson died from cancer at her home in Silver Springs, Maryland, on April 14, 1964, at age fifty-six, less than two years after the publication of *Silent Spring.* Many prominent government officials, scientists and conservationists attended her funeral held in Washington's famed National Cathedral. On the floor of the United States Senate, Senator Abraham Ribicoff of Connecticut paid a tribute to "this gentle lady who aroused people everywhere to be concerned with one of the most significant problems of mid-twentieth century life."

"There is no question," said one government expert on natural resources, "that *Silent Spring* prompted the federal government to take action against water and air pollution—as well as against pesticides—several years before it otherwise would have moved."

On June 27, 1970, Secretary of the Interior Walter Hickel dedicated a national wildlife refuge to the late Rachel Carson in Wells, Maine. In unveiling a plaque at the woodland site, overlooking a picturesque tidal marsh bordering the ocean, Hickel pointed out that the author of *Silent Spring* has been credited by many with starting the age of ecology in the United States. Significantly, this bird refuge is located near the late biologist's summer home at West Southport.

Frank Graham, Jr., her biographer, has said of the author of *Silent Spring*, "If America ever chooses to adopt a sane, coordinated conservation policy—an *environmental* policy —a great deal of the credit must go to Rachel Carson. She did more than alert the public to a difficult and critical problem. She uncovered and pointed out publicly for the first time, even to many scientists, the facts which link modern contaminants to all parts of the environment. There are no separate environmental problems, Rachel Carson insisted. She synthesized the issue—for the scientists, the public and the government."

14
Joseph Wood Krutch

For almost thirty years, until 1952, Joseph Wood Krutch (pronounced Krewch) was an outstanding literary and Broadway drama critic. From that year until the end of his life at age seventy-six, on May 22, 1970, he was a leading field naturalist and conservationist in the American Southwest. On both subjects he wrote with great distinction, but his books and articles on nature brought him national recognition. The last eighteen years of his life he spent in Tucson, Arizona, where he went to escape the confines of big city life and to recover from a respiratory ailment. During this time conservation and the protection of our environment became a national concern. They affected the lives of people more deeply than did the Broadway stage, and for this reason we will remember Krutch more for his second career in the West than for his earlier distinguished one in the East.

Many people have wondered how he could have made a smooth transition from theater criticism to preventing the majesty of the Grand Canyon from being ruined by man-made dams. One clue to this mystery was an early Krutch biography of Henry David Thoreau, one of his heroes. Furthermore, his position as Brander Matthews Professor of Dramatic Literature at Columbia University was not earned by covering of glamorous Broadway first nights. It reflected a depth and range of knowledge that went far beyond the stage.

Reflecting his phenomenal intellect, Krutch's work had an overriding theme throughout his long life: the relation of man to the universe. In 1929 Krutch penned a pessimistic book, *The Modern Temper,* which shocked many readers with his statement that: "living is merely a physiological process with only physiological meaning. And . . .it is most satisfactorily conducted by creatures who never feel the need to attempt to give it any other. Ours is a lost cause and there is no place for us in the natural universe." His thesis was that scientific reality was incompatible with the human spirit.

In *The Modern Temper* Krutch foreshadowed his later interest in nature when he asserted that all great civilizations decay eventually and are rejuvenated by primitive people who have "an animal acceptance of life for life's sake." While some reviewers found him unduly pessimistic in this work because of his rejection of moral and aesthetic values for man's past, others hinted that he was on the right track.

Krutch was born in Knoxville, Tennessee, in 1893. After receiving a B.A. from the University of Tennessee, he did graduate work in the humanities at Columbia University to earn both an M.A. and a PhD. degree.

From 1924 through 1952 Joseph Krutch was drama critic for *The Nation*, the liberal weekly journal, and during that period he produced books at the rate of almost one a year. In addition between 1943 and 1952 he taught at Columbia University where he was named Brander Matthews Professor of Dramatic Literature.

After publishing *The Modern Temper* (1929) and several other notable works, including a biography of Edgar Allen Poe, as well as drama anthologies, Krutch began to get the conservation bug. He was thirty-seven when he first read Thoreau's *Walden.* The Concord naturalist's classic and its integrated natural history made an indelible impression on him. The year was 1930 and America was in the first year of the Great Depression.

Almost twenty years later (1948) Krutch published his critical biography of the Massachusetts nature lover, Thoreau, whom he quoted more than any other writer to the end of his life. Krutch began his work with this significant sentence —"The lesson which Henry David Thoreau taught himself and

which he hoped he might teach to others was summed up in one word: 'simplify!'" Rejecting the notion of man as the master of Nature, he liked to cite Thoreau's view of a world "more to be admired than it is to be used." When Krutch and his wife, Marcelle, moved to Arizona in the early 1950s, partly for reasons of health, he confessed that he was motivated intellectually to explore the Southwest because: "I felt that the time had come when I should take a closer look at the part of my universe which neither I nor my fellows had made." He was finally escaping the man-made world that he had described as stifling and doomed in his gloomy book, *The Modern Temper,* written twenty-three years before.

Already swayed by the impact of Thoreau and *Walden*, Krutch finally became "convinced that the natural as opposed to the wholly man-made world had become, for me at least, a necessary part of the context of a Good Life." This motivation to go West, as Horace Greeley had done over a century before, did not come from a "romantic primitive's" motivation to return to the woods as Thoreau had done, but because he knew that a return to nature would be the most rewarding experience of his life.

Then began the happiest and most fruitful part of his career. Since he already regarded himself as an amateur naturalist, he commenced a deeper study of biology and related natural sciences so that he could better understand all aspects of life. As he prowled around his favorite haunts in the Arizona desert and the Grand Canyon, he used what he had learned of botany, biology and geology as frames of reference relating to the whole universe. He wrote about these subjects rather than the random pleasures of nature. He no longer felt himself "alienated from man or the universe," having achieved a measure of reconciliation with his environment.

After moving to the desert, Krutch lived in a simple adobe brick home where he did most of his writing and research on nature. As he explained it: "I didn't come West for its future, or its industry, its growth or its opportunity. I came for three reasons: to get away from New York and the crowds, to get air I could breathe and for the natural beauty of the desert and its wildlife."

A string of memorable books on nature and conservation followed, including *The Desert Year, The Voice of the Desert,* and *The Great Chain of Life.*

From 1952 on, Krutch served as trustee of the Arizona-Sonora Desert Museum in Tucson and was devoted to conservation and educating the public on the value of the disappearing natural world. His writing of nature books was occasionally interspersed with collections of essays and philosophical tomes like *The Measure of Man*, which won the National Book Award for Non Fiction in 1955. Earlier he received the coveted Burroughs Medal for nature writing. His works which emphasized humanism over materialism, foreshadowed the new interest in ecology. He warned about the dangers of the population explosion and criticized the unchecked growth of the suburbs, which he called "affluent slums."

Among his more prominent books in the ecological field were: *A Treasury of Birdlore, More Lives Than One* (his autobiography) and *The Best Nature Writing of Joseph Wood Krutch.* The last of these was published just before his death.

He also received wide acclaim for two conservation-oriented television shows which he starred in and helped to produce. The first was an hour-long tour of the Sonora Desert near Tucson which appeared in 1963. Two years later he demonstrated in "The Grand Canyon" his acute perception of the grandeur and mysteries of nature.

Although Krutch had shown an early interest in science and majored in mathematics as an undergraduate, he eventually came to fear the application of science in the modern world.

"Man's humanity, " he wrote, "is threatened by the almost exclusive technological approach in social, political and philosophical thought."

In his writings he often expressed a longing for an older, more contemplative way of life and a distaste for many of the scientific inventions of the twentieth century. "If you drive a car at 70 miles per hour, you can't reflect or think about anything, you can't do anything but keep the monster under control," he said. "I am afraid this is the metaphor of our society as a whole."

Unfortunately, after his move to Tucson, he became increasingly disillusioned as urban problems began to envelop that city year after year and once again sought escape to a more natural "paradise." He found it in Baja California—that long, narrow peninsula which is part of Mexico, flanked by the blue Pacific on one side and the Gulf of California on the other.

In his later years he spent more and more of his time in Baja, seeking the solitude that became more difficult to find in Tucson. One result of this sojourn was another book, *The Forgotten Peninsula.*

In one of his perceptive essays, written in 1962, Krutch wrote, "one of the most striking aspects of the human condition is the simple fact that we share the earth with a vast number and a vast variety of other living things. We are enormously different in many respects from any of them, but we are like them all in that they, too, are that mysterious thing, 'alive.'"

He was definitely in the vanguard of other conservationists of his time who saw the interrelationships of all living things which he first expounded in his *Great Chain of Life* (1956). Many of the earlier conservationists did not see this relationship, often focusing their interests on a single aspect of nature such as trees or birds. Krutch, however, not only clearly saw this web of life phenomenon but was one of the first to warn us that the universe might one day "consist of man and the whirl of the atoms and the stars. He alone will be alive in an otherwise dead world."

But he was realistic enough to realize in his essay on "Man, Nature and the Universe" that gadflies like himself could not do the job alone. "It may be that the attitude of biologists and the way in which they teach their subject may determine, as much as poets and 'nature writers,' what the earth will be like a century from now."

He felt himself a bit different from most mortal men, especially those who were frustrated and alienated from the world about them. In his autobiography he wrote optimistically: "But from what I believe man actually is, what human life can be, I am certainly not alienated. And that, I think, sets me apart from at least many of those who call themselves both alienated and existentialist."

After submerging himself in the natural beauty of the American West, Krutch said near the end of his life that he thought he saw some "drive toward the 'higher' in the order of nature." This feeling was expressed in his exuberance and glowing wisdom toward the out-of-doors which shone through the pages of his final books and writings.

Three years before his death, Krutch, in a more pessimistic mood, concluded that most systems were hopelessly materialistic, and that the intellectual minority had become "nihilistic, interested chiefly in destruction and violence." The young who rejected both materialism and violence and those of the older generation who became concerned about the status of our environment, began to turn to Krutch for guidance out of a confused world.

Just before his death, Krutch prophesied the dilemma of modern man when he wrote in an Arizona newspaper that the decade of "the Seventies may be the beginning of the end, or the beginning of a new civilization. If it becomes the latter, it will not be because we have walked on the moon and learned how to tinker with the genes of unborn children, but because we have come to realize that wealth, power and even knowledge are not good in themselves but only the instruments of good and evil."

It was not on some college campus in 1970 that this wistfully prophetic man, denouncing an economy of waste, wrote that "the ultimate good was not a rising standard of living but rather a system of values whose end is man." Shortly after his death in mid-1970, the *New York Times* editorialized: "The idea ran through his writings over a period of four decades, coupled with warnings that the reality revealed by science was not necessarily, or even generally, compatible with the human spirit."

His work had a curiously old-fashioned ring—until suddenly the world, which had ignored him in the past, took up his earlier contention that "smog, pollution, the horrors of war" could be eliminated by a technological society if only it had the sense of values to put such objectives above its greater desire for "more speed, more power and more wealth."

In his lifetime Krutch made the harmonious shift from drama critic to naturalist because the same "wildness" that had haunted Thoreau's imagination a century before had in turn fascinated him. He, too, like Thoreau made his mark among the outstanding American nature lovers and conservationists. He had achieved a long-sought goal, in Thoreau's incomparable phrase "to meet myself face-to-face."

15
Ian McHarg

Ian McHarg, a tall, bearded, briery Scot with a bushy handlebar mustache has been called by *Life* magazine an "eloquent and important successor to Rachel Carson.... He has a respected national reputation in the field of ecology." McHarg likens man's presence on earth to that of a "planetary disease." He refers particularly to the actions and reactions of Western man throughout the centuries with respect to his environment.

As chairman of the Department of Landscape Architecture and Regional Planning at the University of Pennsylvania McHarg has acted as a vociferous critic, picking up where the soft-spoken Rachel Carson left off. "The Judean charge to man to go forth and multiply and subdue the earth has all too efficiently been carried out, rendering this earth a deadly blow," McHarg said in a speech which focused on the roots of his concern—that man has sowed the seeds of his own destruction.

McHarg speaks like an express train rushing along the rails at ninety miles an hour and sometimes words get lost in his superfast delivery, which is combined with his Scottish burr. But from his pithy, thought-provoking and entertaining remarks, which have been seen and heard over many network television programs, the listener can pick up such recurrent themes as: "Man has to be considered an epidemic on this earth," and "if we judge man's brain by man's use of it, then one must conclude that the brain is a spinal tumor."

McHarg is particularly incensed at the urban sprawl on the American East Coast and the way in which the once-beautiful landscape is marred by the blight of "hot dog stands, gas stations, diners and subdivision ranchers." He has documented the way we have "selected beautiful rivers for junkyards, and to dispose all that is foul...as though we prefer a dilute of super-dead bacteria in chlorine solution instead of water."

He feels that, although we have not consciously made this choice, the time has come when "we cannot indulge the despoiler (man) any longer. He must be identified for what he is—as one who destroys the inheritance of the living and unborn Americans, an uglifier who is unworthy of the right to look his fellows in the eye, whether he is industrialist, merchant, developer, Christian, Jew or agnostic."

In his set speech, which he gives over and over with some flowery variations, McHarg launches an attack on such beliefs as the auto being "pre-eminent over man and that open spaces in the cities are a positive evil to be eradicated." He sides with nature in land development controversies and makes a strong plea for man to work in harmony with nature instead of at cross-purposes. He is against man's filling of the marshes just to get more useable land, since such areas are indispensable to recharging nature's batteries.

He makes a plea for retaining large open-space areas in our cities, which can be achieved without greatly increasing the time-distance from city center to urban fringe. Although nature would appear to be defenseless against the encroachments of modern urbanization in the spreading megalopolises, in Ian McHarg nature has found a true champion.

Ian McHarg was born in Clydebank, Scotland, on November 20, 1920, a city where many of the world's largest ocean liners were built. During World War II he rose to the rank of major in the British army, commanding parachute troops. After seven years of military service he was discharged in 1946 and emigrated to America where he enrolled at Harvard, receiving a bachelor's degree in landscape architecture in 1949. This was followed by a master's degree in the same field in 1950 and a second master's in city planning in 1951, all from the same institution.

After leaving Harvard, his professional academic career has centered in Philadelphia, where he rose through the ranks at the University of Pennsylvania's Department of Landscape Architecture to the post as chairman. On the non-academic side, he is associated with one of the city's more prestigious architectural firms, Wallace, McHarg, Roberts and Todd, which with others has recently completed a study for the mayor of New York to redesign the entire lower end of Manhattan Island.

The winner of many awards and honors, McHarg is proudest of his ecological studies conducted for the Twin Cities (Minneapolis and St. Paul); Staten Island; Montgomery Co., Pa. (Skippack Reservoir Study); Baltimore (Inner and Green Spring plans) and Washington, D. C. (Comprehensive Landscape Plan).

These studies have firmly established McHarg as one of the leading thinkers and designers who are welding architecture and ecology in order to improve the environment —before it is too late.

Although McHarg has published only one book, *Design with Nature* (1969), he had previously established his reputation with some dozen articles in journals and edited volumes, most of them having to do with man and his environment and the ecology of the city. He has also received kudos and publicity from stories in leading newsmagazines and from his many appearances in television documentaries, in which he has deftly used his colorful prose to narrate film showing how man has despoiled his environment.

Lewis Mumford, a leading social and environmental critic of our time says about McHarg's important book, *Design with Nature,* "He who touches it touches a man." This seminal work is one of a few which treats of man in relationship to his total environment

Graphs showing unplanned growth of suburbs and under-utilization of open spaces in a metropolitan area.

Pattern of land consumption in a metropolitan area.

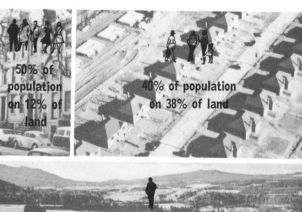

How we use and share the land in a metropolitan area.

—both his fellow creatures and the planet earth they occupy. "All the forces and animate beings that have helped to make man himself what he is."

Some, like Mumford, have likened McHarg's book to the small shelf of environmental classics which include: the ancient Greek Hippocrates' famous medical work on *Airs, Waters and Places* ("the first public recognition that man's life, in health and in sickness, is bound up with the forces of nature"), George P. Marsh's *Man and Nature* and Rachel Carson's *Silent Spring*. Wolf von Eckardt, the Architecture Editor of *The Saturday Review,* has said of McHarg's volume, "It may well be one of the most important books of the century, a turning point in man's view and treatment of his environment."

McHarg states over and over that nature must be treated as an ally and friend and not as an enemy. He has pointed out man's pollution and destruction of nature despite the early warnings so often given us.

He, too, tells of the terrible results of man's misuse of chemical poisons in pesticides, herbicides and detergents as broached earlier in Rachel Carson's *Silent Spring*. But he has gone a step further than merely recapitulating man's destruction of his planet. With his planner's mind, he makes a distinct contribution to a new kind of conservation.

McHarg shows how new knowledge can be applied to caring for our natural areas, the rivers, lakes, swamps and forests, so that man can re-establish human norms and choose sites for new urban settlements in order that the aesthetic quality of life may also be enhanced.

A new type of ecological thinker, he mixes scientific insight with visionary planning and design ability. Where earlier conservationists taking a cautious view have advised leaving nature in charge, McHarg boldly proposes that man attempt to modify and improve upon his environment while cooperating with nature.

"McHarg's emphasis is not on either design or nature by itself," says Mumford in an introduction to his book. His emphasis is rather "on the preposition *with*, which implies human cooperation and biological partnership. He seeks, not arbitrarily to impose design, but to use to the fullest the potentialities—and with them, necessarily, the restrictive conditions—that nature offers."

McHarg knows that man has it in his power to work with nature, using his mind, which is a part of nature. McHarg's work is a call to action but does not offer easy or quick solutions to the problem of pollution.

"Here are the foundations. . . .that will replace the polluted, bulldozed, machine-dominated, dehumanized, explosion-threatened world that is even now disintegrating and disappearing before our eyes," writes Mumford.

McHarg paints a future of material progress and human happiness which can become reality if we treat nature with the respect that enlightened ecological thought demands. He uses appropriate photographic illustrations to show his readers the before and after effects of the proper planning of man's creations to harmonize with nature.

In a recent Smithsonian Institution report entitled *The Fitness of Man's Environment,* McHarg wrote that man must learn and understand the value system of nature before he can properly cope with the problems of making our machines and buildings fit in with nature's pattern in the creative process. McHarg often got bogged down, however, in dense academic jargon like—"The measure of success in this process, in terms of the biosphere, is the accumulation of negentropy in physical systems and eco-systems, the evolution of apperception or consciousness, and the extension of symbioses—all of which might well be described as creation." But McHarg would usually make his point later on with simpler language or in satire, as in this version of the American dream: "Give us your poor and oppressed, and we will give them Harlem and the Lower East Side, Bedford-Stuyvesant, the South side of Chicago and the North of Philadelphia—or if they are lucky, Levittown."

His satire and bitter sarcasm have served him well in delivering his written and verbal messages around the country through various media during the past ten years. McHarg is a true forerunner of the new breed of modern day ecologists who relate their original professional training to the larger effort to balance all the aspects of man and his physical environment. McHarg's professional discipline is landscape architecture and regional planning, from which his interest in ecology has evolved. He has been a pioneer in welding the two together. Future ecologists will undoubtedly follow in his footsteps with more concrete plans, designing new ways to help man get right with nature.

He has served as a bell-ringer, warning of further chaos if man doesn't do something soon to reverse his current trend of haphazard development and misuse of natural resources. If there is any one major criticism of McHarg's contribution to the ecological revolution to date, it is that his approach to solutions is sometimes vague, leaving his readers and listeners in a state of semi-confusion.

But in his role as a landscape architect and urban planner McHarg has added a new ingredient to the ecological movement. Like others, he works to reverse man's present wasteful path but in negotiating a new contract with nature he sees a larger role for creative planning and the design functions of society. Only then will man-made problems yield and the quality of our environment be improved.

16
Paul Ehrlich

The man of our times who argues every day that the chief cause of pollution is not the automobile but too many people is a rawboned six-foot, two-inch tall biologist in his mid-thirties who prefers to wear long sideburns and a short haircut. Dr. Paul Ehrlich is worried that his teen-aged daughter might not live through some kind of man-made holocaust in the next decade—unless we do something drastic to prevent man from converting this once beautiful blue planet into a global wasteland surrounded by dead oceans and seas.

This dire prophecy is repeated over and over again for his vast audiences, which consist of readers of his best-selling books, college groups and television viewers. Ehrlich is booked for personal appearances over a year in advance and he receives over three dozen requests every day.

He travels over 80,000 miles a year and tries to keep up an eighteen-hour-day pace, while playing the self-appointed role of a modern day Paul Revere, alerting Americans to their possible impending doom if the present course of environmental destruction is allowed to continue.

Ehrlich admits he is an alarmist, since he is concerned that over three-quarters of his fellow men go to bed hungry every night. He is a man in a hurry and speaks in a rapid fire rumble as he paces around his classroom, office or lecture platform, usually wearing a wrinkled suit that always seems to need a pressing.

It took one million years, he says, to double the world's population from 2.5 million people to 5 million by 6000 B.C. In the early 1970s he tells us, we are now pressing 4 billion and the doubling time is only thirty-seven years. If we keep on at this alarming rate, Ehrlich points out that we could only export to the stars one day's increase in the population each year. The cost would be prohibitive. Remember it costs us *only* $250 million to send two men to the moon for a five-day round trip!

Even conceding that one day we might reach the planets and populate them, he fears that all the "material in the visible universe would soon be converted into people and the sphere of people would be expanding outward at the speed of light." For those who believe that war might be a leveling factor, he coldly calculates that "all the battle deaths suffered by Americans in all wars (which amount to over 600,000) have been more than made up by births on this globe in the last three days."

He is pessimistic about the so-called "Green Revolution" which some economists and agronomists predict will save us all. Rather, he feels the earth and the crops will soon turn brown because of the lack of genetic variability in the crops.

He is not afraid to point his finger at the men whom he calls "environmental villains" or "ecological Uncle Toms" who serve the Establishment in Washington. His particular pet peeves are the physicists who rule the scientific roost in this country and control the leading technologically-oriented public and private organizations like the Atomic Energy Commission and the National Academy of Sciences.

Ehrlich came naturally to biology, beginning with his childhood in Philadelphia, where he was born in 1933. He was fascinated by butterflies, writing a book on the subject in 1961. He pursued his interest by majoring in biology at the University of Pennsylvania in 1953. From there he went west to the University of Kansas, where he received his M.A. and PhD. degrees. After a short stint at the Chicago Academy of Sciences, he joined the faculty of Stanford University, where he served for three years as director of graduate study for the Department of Biological Sciences.

He is currently a full professor of biology at that institution. Ehrlich considers himself a scientist and researcher by profession and a missionary for the ecology movement only by reluctant choice. He presents a rare combination of natural eloquence and articulate expertise in the new field of population biology —the study of how species naturally control their growth and size.

From the beginning of his teaching career, he found himself both inspiring and vexing his students. As a result of his teaching a recent full-house class consisting of some 700 undergraduates, many of these students became missionaries for his cause. They spread the word to others about their professor's theories. Ehrlich then found himself making presentations off campus and being interviewed around the San Francisco Bay area.

On one of those occasions he so impressed the head of America's most prestigious conservation organization, David Brower of the Sierra Club, that the two made plans for a book to be written by Ehrlich as part of the Sierra Club's special conservation paperback series. Ehrlich worked every night for three weeks to produce *The Population Bomb*. This book immediately became a runaway best seller and was the most widely publicized

The more cross-connecting links in an ecosystem the more chances that system has to survive changes.

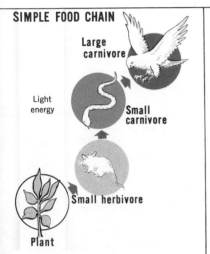
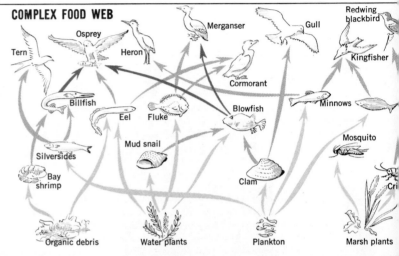

"survival" book of our times. It has gone through twelve printings since its publication in May 1968 and has sold over 2 million copies. This book skyrocketed Ehrlich into sudden prominence and he was much in demand for radio and television shows. He was able to serve as a one-man lobbyist for his dogma that time is running out for man on this overcrowded planet unless some drastic steps are taken to lower the birth rate. Ehrlich found that he was getting through to people who might not otherwise have read his book, or any other for that matter, after two appearances on the popular network "Tonight" show, which drew some of the heaviest viewer responses in that show's history. He has a flare for writing and speaking in simple, down-to-earth language that everyone can understand.

Ehrlich's major thesis in *The Population Bomb* is that the human population of the planet is "about five times too large" and we're managing to support all those people—at today's level of misery—only by spending our capital, burning our fossil fuels, dispersing our mineral resources and turning our fresh water into salt water. "We have not only overpopulated, but overstretched our environment," he says.

"We are poisoning the ecological systems of the earth upon which we are ultimately dependent for all our food, for all our oxygen and for all our waste disposal." He points out that complex environmental systems or "ecosystems," as they are called, are made up of many different kinds of animals, plants and organisms, which we are killing off at a rapid rate.

What we don't realize is that when food webs become simpler with fewer cross-connecting links they become unstable. It is therefore necessary to keep them in their proper balance if we are to avoid catastrophe. Example: by eliminating competitors, insect pests, scavengers — and with them species of wildlife like the California condor, the peregrine falcon and the brown pelican — we reduce the complexity of the systems upon which our very existence depends.

In his book, *The Population Bomb,* Ehrlich propounds, among other things, eleven of mankind's inalienable rights that he feels are worth preserving. These include: (1) the right to eat well; (2) the right to drink pure water; (3) the right to breathe clean air; (4) the right to decent, uncrowded shelter; (5) the right to enjoy natural beauty; (6) the right to avoid regimentation; (7) the right to avoid pesticide poisoning; (8) the right to freedom from thermonuclear war; (9) the right to limit families; (10) the right to educate our children; and (11) the right to have grandchildren.

In subsequent articles written on the same theme, Ehrlich warns those who take comfort in the possibility that the population explosion in this country will start to level off by the end of the century (the 1970 census indicators show that the United States may grow by "only" 75 million people by the year 2000). What they are not taking into consideration is one important factor, namely that each American has roughly fifty times the negative impact on the earth's life-support systems as the average citizen of India. Ehrlich has pointed out that this means that 75 million more Americans are actually equal to 3.7 billion Indians in terms of eco-system destruction. This ratio is based on our ability to pollute the air and water faster than the more primitive cultures.

Therefore he concludes that population growth in this nation "is more serious than population growth in the underdeveloped countries." For this reason we must make great efforts to "change our life style so as to reduce our per capita impact on our environment, *and* to control our population, or disaster will overtake us."

He further warns us, at the risk of being labeled a prophet of doom, that "ecological considerations indicate that only fifty million Americans, living as they do today, could eventually destroy the planet."

He foresees major problems for our children as 280 million Americans struggle to survive into the twenty-first century and predicts that "even with luck, we are doomed to continued population growth until at least 2045"—when the projected population size of the USA will be well over 300 million. This is hardly a pleasant prospect, he concludes, for a nation now failing to provide properly for 205 million people.

By means of his multi-media approach Ehrlich has helped educate the public on the hidden effects of overpopulation and pollution. He cites an example of what he calls a *synergism* (the constructive or destructive interaction of two or more factors that yield a total effect greater than would occur if the factors operated independently). His illustration of an environmental health synergism described the deleterious effects of the interaction of sulfur dioxide from coal-burning power plants and asbestos particles from automobile brake linings, which together induce lung cancer. The sulfur dioxide interferes with the process by which foreign particles are expelled from the lungs, which in turn, increases the residence time in the lungs of carcinogenic asbestos, hence the increase in the chances of man's contracting this deadly malady.

Ehrlich is also not afraid to leave the safe confines of the laboratory and enter into the political arena. He is president of a new group called Zero Population Growth (ZPG) that is dedicated to stopping the population escalation and environmental deterioration in this country through meaningful political action.

In every speech he makes, Ehrlich attacks the national leadership of this country regardless of party affiliation for its ecological ignorance and irresponsibility, sometimes labeling the president and other responsible government officials as simple "boobs." His predominantly young audiences respond favorably to his blunt talks, and Ehrlich has criss-crossed the country on the college circuit, aiming to mobilize the campuses to play a leading role in environmental activism.

Like Ralph Nader, the self-appointed spokesman for the consumer protectionist movement, Ehrlich has assumed the role as chief propagandist for the emerging population control movement, since he deeply believes that all other forms of pollution—water, air, noise, waste, etc.—are all secondary to the main form of pollution, people pollution.

Ehrlich's hectic pace has played havoc with his home life. Furthermore, he has had to defer work on two basic biological texts that emphasize ecological considerations, a dual project that has been incubating for several years. "Ecology has been largely ignored in biology teaching over the past few decades," Ehrlich has stated, "but it's certainly *the* area of biology that should be emphasized today."

His new-found, public career has forced him to give up his favorite form of recreation —piloting his own small airplane. Now most of his traveling is done by commercial airliners, on his way to give another talk to alert people on the dangers facing the sky above and the earth below.

He regrets that he can't spend more time with his wife, Anne (who co-authored his other book, *Population, Resources and Environment*), and his only child, a teen-aged daughter, Lisa. Ehrlich often makes the point that "population control starts at home" and tells his audiences that he is married, has only one child and has had a vasectomy, a form of sterilization for the male which results from a simple, painless operation.

The author of over ninety scientific articles, Ehrlich published a paperback book in January 1971 entitled: *How to be a Survivor*. In this latest co-authored venture he presented a no-holds barred blueprint for human survival on this shrinking planet. In it he proposed a crash program of radical changes that will be needed in order to meet the environmental crisis of our times. Since he merely outlined general paths to solutions in his first book on population, this more specific book was a natural sequel. It propounds the platform of his ZPG organization and has become a sought-for bible for that group of concerned citizens.

As a lobbying group, Ehrlich's Zero Population Growth organization presses for legislation to implement a far-reaching birth control program, the repeal of archaic legislation that runs counter to these objectives, and for more research into population problems and into better methods of human contraception. In addition, ZPG has worked for tax laws that, instead of offering incentives for having more children, will provide tax incentives for smaller families and will emphasize the need for population control.

His ZPG members pledge that they will limit their offspring to no more than two children. "Anything beyond two children is irresponsible, suicidal," says Ehrlich. So infectious has this Ehrlich-spawned baby become that, by the spring of 1970, 102 chapters were formed in thirty states. This young organization, which was formed in March 1969, is now acquiring new members at the rate of over 500 a week.

ZPG, whose headquarters are located in a charming suburb of Los Altos, California, not far from the Stanford campus, is presently working hard to elect candidates in 1972 who will help solve our problems and to defeat those who don't understand the pollution problems facing man or who represent special interest groups. ZPG also organizes picketing at hospitals with what Ehrlich calls "antique sterilization policies" and works for abortion reform, smog control and other ways to improve our environment.

Both ZPG and Ehrlich are convinced that the rising ecology movement is not a fad. Theirs is a long-term mission that will not soon disappear, since the underlying problems continue to escalate daily.

Although Ehrlich denies that he is any sort of an ecology hero, he happily acknowledges the emergence of the new youth movement based on the population-pollution issue, but warns that it must become more than a youth movement. "If anything is going to bring us all together, poor, rich, black, white, young, old, this has got to be it," he says emphatically in an appeal to older adults to join in the movement.

He points out that "affluence and effluence" go hand-in-hand. "We Americans, comprising only six percent of the world's people consume thirty percent of the world's available resources each year," he says. "So you can see that each American child puts far more strain on the world environment than each Asian child. We've got to put our own house in order before we start telling Asians and Africans what to do."

He believes that the two groups of people most distrusted—the very old and the very young—could form a great force for change, if they could come together and put pressure on the middle-aged adult group that is responsible for so much destruction of our environment. Ehrlich is heartened by the young's disdain for material things, their fascination for nature, and interest in what might be called an ecological way of life.

He is careful to point out that the environment issue is not a "safe cop-out" for the "real" issue of poverty, race and war. "Your cause is a lost cause without population control," he tells his youthful audiences, "and race, war, poverty and environment are really part and parcel of the same big mess."

Ehrlich believes that the expanding ecological revolution is going to generate "a lot of civil disobedience, similar to what we saw in the early days of civil rights: demonstrations, picketing, sit-ins. I think we will soon

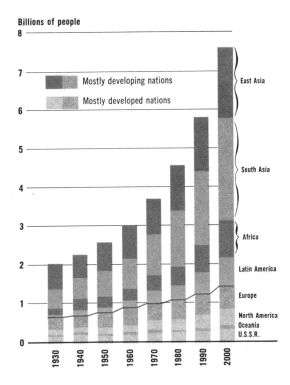

Projected growth of world population based on the United Nations "constant fertility" projection.

begin to see boycotting of the automobile industry, the big oil companies, the utilities and so on. Among other things, people are just going to stop paying their bills. One clue that we're making real progress will come when politicians start telling it like it is and to hell with the consequences."

In his recent book, *Population, Resources and Environment*, Ehrlich sums up his assessment: "Spaceship Earth is now filled to capacity or beyond and running out of food. And yet, people traveling first class are, without thinking, demolishing the ship's already overstrained life-support systems . . . . Thermonuclear bombs, poison gases, and super germs are being manufactured and stockpiled. . . . But, unaware that there is no one at the controls of their ship, many of the passengers ignore the chaos or view it with cheerful optimism, convinced that everything will turn out all right."

Environmental Organizations and Periodicals

American Forests (Monthly)
The American Forestry Association
919 Seventeenth Street, N.W.
Washington, D. C. 20006

Audubon (Bi-monthly)
Conservation Guide (Twice monthly)
National Audubon Society
1130 Fifth Avenue
New York, N. Y. 10028

CF Letter (Monthly)
The Conservation Foundation
1717 Massachusetts Avenue, N.W.
Washington, D. C. 20036

Council on Environmental Quality
722 Jackson Place, N.W.
Washington, D. C. 20006

The Defenders of Wildlife
2000 N Street, N.W.
Washington, D. C. 20036

Environment (10 issues per year)
The Scientists Institute for Public Information
30 East 68th Street
New York, N. Y. 10021

Environmental Action Bulletin (Weekly)
Environmental Action
1346 Connecticut Avenue, N.W.
Washington, D. C. 20036

Environmental Protection Agency
1626 K Street, N.W.
Washington, D. C. 20460

Not Man Apart (Monthly)
Friends of the Earth
451 Pacific Avenue
San Francisco, California 94133

The Garden Club of America
598 Madison Avenue
New York, N. Y. 10022

The Izaak Walton Magazine (Monthly)
The Izaak Walton League of America
1326 Waukegan Road
Glenview, Illinois 60025

National Parks
& Conservation Magazine (Monthly)
National Parks & Conservation Association
1701 Eighteenth Street, N.W.
Washington, D.C. 20009

Parks & Recreation (Monthly)
National Recreation & Park Association
1700 Pennsylvania Avenue, N.W.
Washington, D. C. 20006

National Wildlife (Bi-monthly)
Conservation News (Twice monthly)
National Wildlife Federation
1412 Sixteenth Street, N.W.
Washington, D. C. 20036

Natural History (10 issues per year)
The American Museum of Natural History
Central Park West at 79th Street
New York, N. Y. 10024

The Nature Conservancy News
The Nature Conservancy
2039 K Street, N.W.
Washington, D. C. 20006

Sierra Club Bulletin (10 issues per year)
Sierra Club
1050 Mills Tower
San Francisco, California 94104

The Living Wilderness (Quarterly)
The Wilderness Society
729 Fifteenth Street, N.W.
Washington, D. C. 20005

Zero Population Growth
330 Second Street
Los Altos, California 94022

Selected Bibliography

American Association of University Women, *A Resource Guide on Pollution Control.* AAUW, Washington, D. C., 1970.

Carson, Rachel, *Silent Spring.* Houghton Mifflin Company, Boston, 1962.

Clepper, H. E., ed., Natural Resources Council, *Leaders of American Conservation.* The Ronald Press Company, New York, 1971.

Cutright, Paul R., *Theodore Roosevelt, the Naturalist.* Harper and Row, New York, 1956.

DeBell, Garrett, ed., *The Environmental Handbook.* Friends of the Earth/Ballantine Books, Inc., New York, 1970.

Derleth, August, *Concord Rebel: A Life of Henry David Thoreau.* Chilton Books, Philadelphia, 1962.

Douglas, William O., *Wilderness Bill of Rights.* Little, Brown, and Company, Boston, 1965.

Ehrlich, Dr. Paul R., *The Population Bomb.* Ballantine Books, Inc., New York, 1969.

Environmental Action Committee, *Earth Day: The Beginning.* Bantam Books, Inc., New York, 1970.

Environmental Action Committee, *Earth Tool Kit.* Pocket Books, New York.

Farb, Peter, *Ecology.* Life Nature Library, Time-Life Books, New York, 1963.

Ford, Alice, *John J. Audubon.* University of Oklahoma Press, Norman, 1965.

Graham, Frank, Jr., *Man's Dominion: A History of Conservation In America.* M. Evans and Company, New York, 1971.

Graham, Frank, Jr., *Since Silent Spring.* Houghton Mifflin Company, Boston, 1970.

Grossman, Mary L. and Hamlet, John N., *Our Vanishing Wilderness.* Grosset and Dunlap, Inc., New York, 1969.

Holt, Rackham, *George Washington Carver: An American Biography.* Doubleday & Company, Inc., Garden City, 1963.

Krutch, Joseph Wood, *The Great Chain of Life.* Houghton Mifflin Company, Boston, 1957.

Marsh, George Perkins, *Man and Nature,* 1874; republished as *Earth as Modified by Human Action,* 1882. Modern edition, Harvard University Press, Cambridge, 1965.

Marx, Wesley, *The Frail Ocean.* Ballantine Books, Inc., New York, 1970.

McGeary, M. Nelson, *Gifford Pinchot: Forester-Politician.* Princeton University Press, Princeton, 1960.

McHarg, Ian, *Design With Nature.* Natural History Press, New York, 1969.

Michener, James, *The Quality of Life.* J. B. Lippincott Company, Philadelphia, 1970.

Milne, Lorus & Margery, *Life on Earth.* Crown Publishers, Inc., New York, 1970.

Mitchell, John G., ed., *Ecotactics: The Sierra Club Handbook for Environmental Activities.* Pocket Books, New York, 1970.

Pinkett, Harold T., *Gifford Pinchot: Private & Public Forester.* University of Illinois Press, Urbana, 1970.

Portola Institute, *Whole Earth Catalog.* Berkeley, 1969.

Reich, Charles, *The Greening of America: The Coming of a New Consciousness & the Rebirth of a Future.* Random House, Inc., New York, 1970.

Reinow, Robert and Leona, *Moment In The Sun.* Sierra Club/Ballantine Books, Inc., New York, 1969.

Saltonstall, Richard, Jr., *Your Environment & What You Can Do about it.* Walker & Company, New York.

Shepard, Paul & McKinley, Daniel, eds., *Subversive Science Essays Towards an Ecology of Man.* Houghton Mifflin Company, Boston, 1968.

Storer, John H., *The Web of Life.* Signet Books, New York, 1966.

Swatek, Paul, *The Users' Guide to the Protection of the Environment.* Friends of the Earth/Ballantine Books, Inc., New York, 1970.

Thoreau, Henry D., *Walden, and Other Writings.* Krutch, Joseph Wood, ed., Bantam Books, Inc., New York.

Thoreau, Henry D., *Journal of Henry D. Thoreau.* Torrey, Bradford & Allen, eds., Dover Publications, Inc., New York, 1906.

Udall, Stewart, *The Quiet Crisis.* Holt, Rinehart, & Winston, Inc., New York, 1963.

Index

Activism, environmental, 88
Adams, John Quincy, 31
"Affluent slums," 78
Agassiz, Louis, 26
"Agricultural-chemical complex," 73
American Association for the Advancement of Science, 37
"American Forests, The," 48
American Ornithologists' Union, 42
American Ornithology, 14, 15, 16, 17, 18
American Philosophical Society, 18
A Report on the Lands of the Arid Regions of the U.S., 36
Arizona-Sonora Desert Museum, 78
A Thousand Mile Walk to the Gulf, 44
Audubon, Capt. Jean, 19
Audubon, John, 21
Audubon, John James, 9, 14, 16, 19-24
Audubon, Victor, 21
Audubon Medal, 12, 74
"Audubon Plumage Bill," 42-43
A Week on the Concord and Merrimack Rivers, 25, 26

Bachman, John, 22
Back-to-nature experiment, 9
Bakewell, Lucy, 19
Ballinger, Richard, 62
Bartram, John, 14
Bartram, William, 14, 15, 17
Bartram homestead, 14, 17
Beebe, William, 57
Benét, Stephen Vincent, 24
Beyond the Hundredth Meridian, 39
Biltmore estate of George Vanderbilt, 60
Birds of America, 21
Birth control program, 88
Black Hills, South Dakota, 59
Block faulting, 45
Borglum, Gutzon, 59
Bradford, Samuel, 15
Bradley, Edwin, 41
Bradley, Guy, 40-43
Bradley, Louis, 41
Breaking New Ground, 64
Brower, David, 85-86
Buchheister, Carl, 74
Bureau of Ethnology, 38
Burlington, Vermont, 30
Burroughs, John, 9, 35, 50-53, 57
Burroughs Medal, 78

Campbell, Tom, 67, 68
Camping and Tramping with Roosevelt, 52

Cape Sable, Florida, 43
Carson, Rachel, 11, 12, 70-75
Carver, George Washington, 65-69
Carver, Moses, 65
Century, 47
Channing, William Ellery, 26
Chapman, Frank, 40, 57
Chemical poisons, 82
Civilian Conservation Corps, 10
Clean Air Bill, 11
Cleveland, Grover, 48
Commercial Fisheries Bureau, 11
Concord, Massachusetts, 26
"Conservation," 54, 55
Conservation Congress, 56
Crane, Caroline, 30
Crop rotation, 65

David, Jacques Louis, 21
Department of Landscape Architecture and Regional Planning, 80, 81
Department of the Interior, 54
Design with Nature, 81
Detergents, banning of, 12
De Voto, Bernard, 36
"Dust Bowl," 39

Earth Day, 8
Earth Modified by Human Action, The, 32
East Cape Sable, Florida, 40
Ecology, science of, 8
Ecology, historical development of, 9
Eco-system, 86-87
Eddy, William, 50
Edge of the Sea, The, 70
Edison, Thomas, 69
Ehrlich, Paul, 84-89
Eisenhower administration, 11
Emerson, Ralph Waldo, 25, 29, 50
Environmental Protection Agency (EPA), 11
Erosion, steam, 47
Erosions, glacial, 45
Everglades National Park, 43
Experiment, back-to-nature, 25
Experimental farm, Carver's, 67-68
Explorations of the Colorado River of the West, 38

"Feather merchants," 41-42
Fitness of Man's Environment, The, 83
Food chain, 9, 71-73
Forest preservation, 60-61
Forest preserves, national, 47-48
"Forest Reservations and National Parks," 48
Forest Service, 61
Forgotten Peninsula, The, 78

Game wardens in Florida, 42
Garbage, human-generated, 9
Glaciation, 47
Graham, Frank, Jr., 75
Great Chain of Life, 79
"Green Revolution," 84

Harris, Edward, 21
Havell, Robert, 21
Hetch-Hetchy Valley, Calif., 49
Hickel, Walter, 75
How to be a Survivor, 88
Hughes, Charles Evans, 43
Hutchings Hotel, 46, 47
Hydrocarbon products, 70, 71

Ickes, Harold, 10
Illinois State Natural History Society, 37
Indians, Chickasaw, 17
Industrial Revolution, effects of, 28
Inland Waterways Commission, 56
Interstate Commerce Commission, 55
In The Hemlocks, 51
Iowa State Agricultural College, 66, 69
Irrigation Survey, 39

Jefferson, Thomas, 17
John Burroughs Memorial Association, 53
Johnson, Robert U., 47

Kennedy-Johnson administration, 11
Key West, 40
Key West shooting, 40
Krutch, Joseph Wood, 76-79

Lacey, John, 48
La Follette, Robert, 59
Lake Erie crisis, 10
Lee, Alice Hathaway, 54
Lehman, George, 21
Lewis, Meriwether, 17
Life Without Principles, 28
Lone Pine, California, 46

MacGillivray, William, 21
Man and Nature, 30, 32, 35
Marsh, George Perkins, 9, 30-35
Marsh, John, 30
Mason, Joseph, 20
McGee, W. J., 60
McHarg, Ian, 80-83
McLeod, Columbus G., 40
Mill Grove farm, 19, 24
Millinery trade, New York, 41, 42
Modern Temper, The, 76
Monroe County, Florida, 43

92

Monuments, national, 49
Mount Whitney, Calif., 46
Movable school, 68
Muir, John, 9, 35, 44-49, 52, 62
Muir Woods, Calif., 49
Mumford, Lewis, 30, 81-82
Mycology, 66

Nader, Ralph, 88
National Audubon Society, 10, 14, 24, 40, 42, 43
National Conservation Commission, 56
National Oceanic and Atmospheric Administration (NOAA), 11
National parks, 57
National Wildlife Federation, 10
Natural Resources, proposed Department of, 12
Nature, 25
Nixon, Richard M., 11

Oakley Plantation, 20
Ord, George, 17
Ornithological Biographies, 18
Ornithological Biography, 23
Ornithologist, 16

Parks, national, 49
Pearson, T. Gilbert, 42
Pesticides, 9, 70-73
Pinchot, Gifford, 9, 35, 55, 58, 60-64
"Pinchot Roads," 64
Plants as Modified by Man, 66
Plunder, forest, 48
Pollution, 9, 82, 84, 87, 88
Population Bomb, The, 85-87
Population explosion, 78
Population, Resources and Environment, 88, 89
Powell, John Wesley, 36-39
Power Survey Board, 63
President's Science Advisory Committee on Pesticides, 74
Price, Overton, 60
Primer of Forestry, 61
Public Lands Committee, 48

Quiet Crisis, The, 33

Reclamation Act (1902), 39, 56
Ree's Encyclopedia, 15
Ribicoff, Abraham, 75
"Riverby," 51
"Robber barons," 55, 56, 60
Robin, Jean, 19
Robin, Mademoiselle, 19
Roosevelt, Edith Kermit, 54
Roosevelt, Franklin D., 10, 69
Roosevelt, Theodore, 9, 49, 52, 54-59, 60, 61, 68
Roosevelt Dam, 56

Sagamore Hill, 57
Sage of Walden, 28
Sargent, Charles, 48
Schurz, Carl, 36
Sea Around Us, The, 70
Seed fertilization, 69
Sewage disposal, San Diego, 12
Seymour, Lucy, 65
Sharp, Dallas Lore, 53
Sierra Club, 10, 49
Sierra Nevada Mountains, 47
Silent Spring, 10, 70
Silviculture, 60
Simpson College, 66
"Slabsides," 52
Smith, Tom, 40
Smith, Walter, 40
Smithsonian Institution, 31, 38
Society of Artists, 17-18
Soil bank conservation, 68
Soil Conservation Service, 10
Sprague, Isaac, 21
State Emergency Relief Board, 64
Stegner, Wallace, 39
Stevenson, Adlai, 13
Strentzel, Louie Wanda, 44
Synergism, 87

Taft, William Howard, 58, 62
Taylor, Zachary, 31
Tennessee Valley Authority (TVA), 10
Thoreau, Henry David, 9, 25-29, 50, 53, 76-77
Thoreauvians, 26
Training of a Forester, The, 61
Travels, 15

Truman, Harry, 24, 69
Tuskegee Normal and Industrial Institute, 66

Udall, Stewart, 11, 33, 35, 49, 61
Under the Sea Wind, 70
U.S. Fish and Wildlife Service, 70
U.S. Forest Service, 57
U.S. Geological Survey, 36-37, 38
University of Stanford, 84-85
University of Wisconsin, 44
Urban planning, 81
Urban sprawl, 80
Uses of the peanut, 69

Viviparous Quadrupeds of North America, The, 22
von Eckardt, Wolf, 82

Wake Robin, 51
Walden, 25, 26, 28
Walden Pond, 25
Wallace, Henry, 69
Wallace, McHarg, Roberts and Todd, 81
Washington, Booker T., 66, 67, 68
Watkins, "Aunt" Mariah, 65
Weather Bureau, 11
Weavers, oppressed, 14
White, E. B., 71
White, Theodore, 10
Whitman, Walt, 50
Whitney, Josiah, 45, 46-47
Wilson, Alexander, 14-18, 20
Woodstock, Vermont, 30
"Wyoming doctrine," 39

Yosemite, The, 45, 47
Yosemite Guidebook, 46
Yosemite (Inyo) earthquake, 45-46
Yosemite National Park, 47
Yosemite Valley, Calif., 44, 45, 46

Zero Population Growth (ZPG), 87, 88

Credits

American Museum of Natural History: pages 52, 55. Brown Brothers: pages 34 (center), 59 (top), 62. Columbia University Libraries: page 29 (bottom). Doubleday & Company, Inc.: page 68. Harvard College Library: page 49. Houghton Mifflin Company: page 71. Alfred A. Knopf, Inc.: page 47. Library of Congress: page 29 (top). Montgomery County (Pa.) Tourist Bureau: page 21. National Aeronautics and Space Administration: page 8. National Archives: page 39. National Audubon Society: page 75. National Film Board of Canada: page 81 (bottom). National Park Service: page 69. New-York Historical Society: pages 23, 34 (top). New York Public Library, Rare Book Division: page 18. Potomac Planning Task Force, American Institute of Architects: page 81 (graphs). Theodore Roosevelt Association: page 59 (center and bottom). Sierra Club: page 48. Soil Conservation Service: page 34 (bottom).